Spatial Statistics
Illustrated

SPATIAL STATISTICS ILLUSTRATED

WRITTEN BY
Lauren Bennett, PhD
Flora Vale

ILLUSTRATED BY
Flora Vale

Esri Press
REDLANDS | CALIFORNIA

To everyone who has attended our workshops over the years. Your questions and nods and smiles and feedback helped us write every page of this book.

Contents

Foreword

If you've ever attended a workshop by Lauren Bennett and Flora Vale, you already know how talented they are at presenting complex topics in a way that is fun and fully accessible to *literally* anyone (from beginner to expert). This book perfectly summarizes and extends their spatial analysis and data science presentations, providing a tangible resource that will help you put the ideas and methods into practice.

I programmed the original spatial statistics tools in 2004 to complement *The Esri Guide to GIS Analysis, Volume 2*. Today, spatial statistics are an increasingly important component of ArcGIS® software, and the spatial statistics team has grown to include tremendously talented programmers, statisticians, and data scientists. In many ways, this success can be credited to the enthusiasm, wildly popular presentations, and extensive resources developed by Lauren and Flora.

Whenever you learn a new subject (especially a subject as extensive as spatial statistics), it helps to have an accurate foundation to hold and organize the technical details and new information that come at you. To accelerate your learning, this book provides that foundational framework. Even if the idea of statistics makes you a bit anxious, you'll find that the writing style and clear explanations quickly put you at ease. For people already familiar with spatial statistics in ArcGIS, this book reaffirms all that you know in a way that is truly satisfying and provides you with tools for explaining what you know to others. I especially like the questions at the beginning of each chapter, demonstrating the relevance and broad applicability of the spatial statistics methods. And I absolutely love how the engaging illustrations tell the story of how each method works. I'm sure you will enjoy reading and referencing this book as much as I do.

Lauren Scott Griffin, PhD
Esri, Spatial Analytics, Senior Principal Product Engineer
Author of *The Playground Problem* and contributing author for *The Esri Guide to GIS Analysis, Volume 2*

Introduction

Welcome

There has never been a more exciting time to learn about analysis. Now, more than ever, our colleagues, communities, and leaders understand the importance of making data-driven decisions. The field of data science, with its focus on using algorithms to turn data into information and knowledge, has brought analysis into just about every aspect of our lives. This explosion has happened in part because of our unprecedented access to data and computational power. With so many algorithms and buzzwords floating around, it can be overwhelming to figure out where to start or where to go next. In this book, we want to offer an approachable introduction to a key area of data science that offers huge opportunities when solving the complex problems our world faces: spatial statistics.

Spatial statistics is a vast discipline with countless methods and applications. This book focuses specifically on the spatial statistics and spatial machine learning tools and capabilities available in ArcGIS®. We will do our best to explain and illustrate how each method works in a conceptual way. The point isn't to teach you how to write the algorithms, but rather how to know when to use them, how to apply them, and how to interpret them. We don't all have to be engineers of machine learning algorithms, but we do have to be knowledgeable about how the methods work so that we can use them appropriately. Our goal is to help you learn about some of the fundamental concepts and amazing algorithms that power this kind of spatial, analytical approach.

While we're cautioning that we have to be careful about how we use these methods, we also want to say it's not rocket science and you really are capable of doing it. Whether you're new to analysis, or you've been doing complex analysis for years, we will try to introduce these methods in a way that will feel approachable so that you can use them to solve important problems.

We'll start by exploring what spatial statistics are and the key role that spatial relationships play in how we understand the world around us in chapter 1, "Why spatial is special." Then we'll look at how some of the most common descriptive statistics are adapted and expanded to summarize the spatial characteristics of our data in chapter 2, "Means and medians." Next, we'll turn to cluster analysis, including both statistical and machine learning–based approaches. First, we'll explore a few techniques for automating the detection of natural clusters in our data in chapter 3, "Finding clusters with machine learning." Then, we'll discuss some of the most widely used statistical clustering techniques and how they can be applied to both spatial and spatiotemporal data in chapter 4, "Statistical cluster analysis," and chapter 5, "Spatiotemporal pattern mining." Finally, we'll introduce some of the foundational concepts in regression analysis and show how our models can be improved by incorporating spatial relationships in chapter 6, "Modeling spatial relationships and making predictions."

Welcome to the beautiful world of spatial statistics.

Asking the right question

There are countless methods at our fingertips to solve the complex problems that we face. Whether we use traditional statistical approaches, newer machine learning methods, or dive into deep learning and the broader world of artificial intelligence, the first and arguably most important step in an analysis is figuring out what question we're trying to answer. It is easy to fall into the trap of "I have this data, what analysis should I do?" or "I have this new tool in the toolbox, what data can I stuff into it?" This will rarely lead us to the answers we seek or the solutions to problems we need to solve.

Instead, we must first and foremost think about what problem we're trying to solve, and what questions we need to ask to help solve that problem. Let's take the example of ensuring our city is providing an equitable, high-quality, and accessible public transportation system. There is no single tool or algorithm that will solve this problem.

It's a big problem with a lot of complexities and nuance. But we can break it down into a series of answerable questions:

1. Who is currently using public transportation?

2. Where are the communities with the highest need for public transportation?

3. Where are there gaps in access?

4. Who is most impacted by those gaps?

5. What are the current usage patterns for each type of public transit?

6. Are there areas where high usage may indicate the need for additional stops or lines?

7. Are there areas that are more likely to experience delays or interruptions in service?

8. What is the current distribution of resources, and how does that compare to need?

Suddenly, that big, complex problem has at least some building-block questions that we can answer. Now that we've got these smaller, digestible questions, we can start to figure out which approach will help us answer a particular question. And sometimes that algorithm will be state of the art or cutting edge. But sometimes it will be a method that has been around for decades. We don't solve our problem more effectively just because we use a more complicated algorithm or a buzzier approach.

This shift from focusing on the methods to focusing on the question is critical if we want to do great analysis. It is also critical if we want to make sure we're answering the questions that will lead us to understanding and help us drive meaningful action.

Turning data into information

While data is often a starting point in an analytical workflow, it has been said many times that *data does not equal information*. You can be data rich and **information poor**. You can have a lot of data and know nothing at all. In fact, you might even say that sometimes the more data you have, the less you know.

Imagine you have been handed a spreadsheet with a hundred thousand records on it and asked to synthesize it into some useful information.

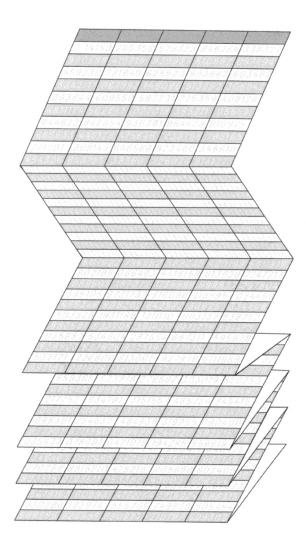

Making sense of this data by just *looking* at all the numbers would be a virtually impossible task. The spreadsheet helps you organize the data but does little to help turn the data into information.

Now imagine that the spreadsheet has a latitude and longitude associated with each of the hundred thousand records. How would you make sense of that spatial data? You could start by plotting those locations on a map…but just as with a spreadsheet, simply displaying your data on a map does not turn it into information.

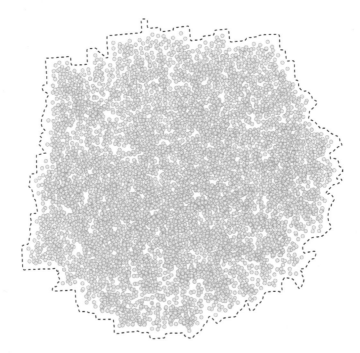

We do get *some* information from putting data on a map. We begin to visualize the distribution, which gives us a little information, but in some ways, the more data you have, the less information you will gain from adding it to a map. We might be able to make sense of 10 points, but not 100,000.

Whenever we look at a map, we immediately start looking for patterns and relationships. Our brains are incredibly powerful statistical machines that are exceptionally good at finding patterns. *So good* at finding patterns, in fact, that sometimes we find patterns even when they're not really there (like faces in toast or bunnies in clouds).

This is where spatial statistics come in. Spatial statistics help us quantify patterns and relationships so that we can feel confident in what we are seeing. This, in turn, allows us to use those patterns, and our understanding of the relationships behind those patterns, to make decisions.

The *point* is (pun intended), we still have a long way to go after putting data on a map before we turn it into real, useful information. Fortunately, this is where spatial statistics can help.

Chapter 1

Why spatial is special

Introduction

Spatial statistics are special. They are not just traditional statistics that we happen to apply to spatial data. They explicitly use some aspect of geography, some notion of space, in the calculation of the statistics. Sometimes this is just incorporating x,y coordinates. Sometimes it is the shape or the orientation of our spatial data. Sometimes it is the distance between locations. But one of the most powerful uses of geography is defining spatial relationships, or what is nearby. These relationships are then used in spatial statistics algorithms to do things like compare local neighborhood averages to global averages to find hot spots, for example. Incorporating nearby features and neighborhoods often provides valuable information that would otherwise be lost in traditional statistical analysis.

This is not to say that traditional statistics are not useful in answering spatial questions. Traditional, nonspatial statistics are appropriate for spatial data if we understand their limitations and ensure that we are not violating underlying assumptions of those methods (more on that later). Some of the methods we'll cover in this book are not inherently spatial. Still, we'll make the case that interpreting these analyses in a spatial context can be very powerful.

Thinking about things like adding spatial variables (for example, variables that represent the distance to roads or water bodies), visualizing and analyzing results spatially, and using results to make spatial decisions demonstrates how these traditional statistical methods can be part of a larger spatial workflow.

The first law of geography

So, why does incorporating space matter so much? That leads us to what we often refer to as the first law of geography, also known as Tobler's law. Tobler's law states that near things are more related than distant things. In other words, things that are closer together are more related than things that are farther apart. This seems intuitive when we say it, but the reality is that many traditional statistical and machine learning–based approaches ignore this reality. In fact, many traditional statistical approaches have underlying assumptions that data is independent. But if Tobler's law holds true, which we know it often does, spatial data (a.k.a. most data) is rarely independent. And those spatial relationships, that dependence between data, is actually a unique and incredibly valuable characteristic of the data.

Ignoring it not only violates statistical assumptions but leaves a lot of potential information on the table.

Spatial statistics build on the concept of spatial relationships in a way that helps us gain a deeper, more valuable understanding of our data. It can help us find patterns that would go unnoticed; it can help us make predictions more accurately, and it can help us make decisions confidently.

Defining spatial relationships

Now we know that spatial statistics use geography. But how?

Any tool or method that uses spatial relationships will require us to define or conceptualize what it means to be neighbors. In other words, we need to define what it means to be related in space (and sometimes even in time). This concept of spatial relationships can feel a bit abstract, but actually it's quite simple. Features in our data can be related in any number of ways. Perhaps two parcels are touching, which makes them neighbors. Perhaps all parcels within a particular drive time of each other are related. Ultimately, every feature will have its own well-defined neighborhood.

To explore this concept, we'll call the feature whose neighborhood we're defining the **focal feature.**

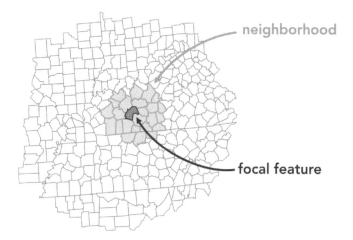

Functionally, spatial relationships are actually represented as a series of weights. Large weights mean lots of influence between features (a strong spatial relationship), whereas small weights represent less influence. For most definitions of spatial relationships, weights can either be binary or continuous. With binary weights, a feature is either included in the neighborhood or it is not. With continuous weights, the magnitude of the weight determines how important the relationship is or how influential the neighboring feature is to the focal feature. There are countless ways to define spatial relationships. Let's explore some of the most common approaches.

Number of neighbors

The number of neighbors method defines a neighborhood using a user-specified number of features (in this case, four) that are closest to the focal feature.

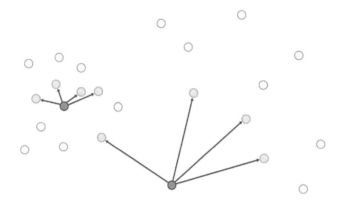

This method is a very powerful, very popular way of defining spatial relationships. Notice how the distances can vary depending on the density of features in the area. Features in dense areas have close neighbors (and therefore smaller neighborhoods), and features in sparse areas have neighbors that are farther away (and therefore much larger neighborhoods). Of course, the number of neighbors is constant throughout each neighborhood, but the size of those neighborhoods can vary quite dramatically. For this reason, this type of neighborhood is often referred to as an adaptive neighborhood.

This approach is also known as k-nearest neighbors, which may seem a little "mathematical" in its name, but k simply represents the specified number of neighbors (in our example, $k = 4$).

Fixed distance

Fixed distance neighborhoods are calculated using a specified Euclidean distance (also known as a straight-line distance or an as-the-crow-flies distance). All features that fall within the specified distance of the focal feature are considered neighbors. Notice how the number of neighbors changes depending on the density of features in the area.

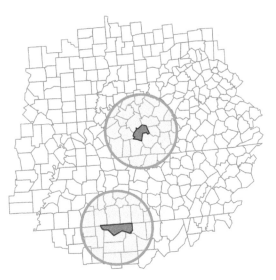

Features in dense areas have many neighbors, and features in sparse areas have far fewer neighbors, using the exact same distance band. For this reason, this type of neighborhood is often referred to as a fixed distance neighborhood, because while the number of neighbors changes, the size of the neighborhood is fixed.

Network distance

Network distance neighborhoods are like fixed distance neighborhoods, but instead of using a Euclidean distance, they use either a network distance or network travel time.

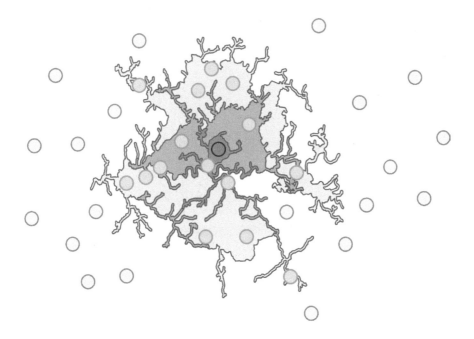

Using a network distance allows our modeling of spatial relationships to be more realistic when looking at human mobility. For instance, two features sitting across a river from each other may have a very short Euclidean distance, but if the only way to get across that river is a bridge 15 miles down the road, then in network distances those features are actually very far apart.

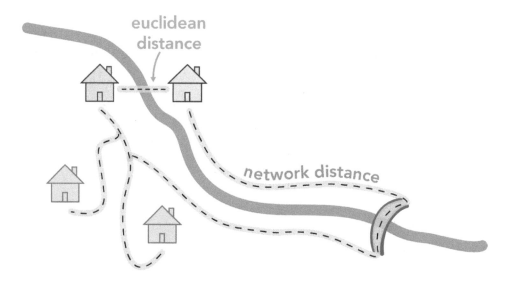

Contiguity

Contiguity neighborhoods define neighbors based on shared boundaries. There are two flavors of contiguity: edges and edges corners. Contiguity edges defines a neighborhood using all features that share an edge with the focal feature.

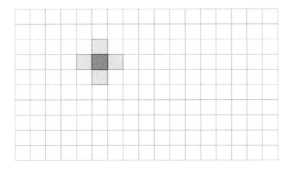

Contiguity edges corners defines a neighborhood using all features that share an edge or corner.

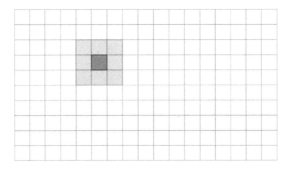

The contiguity relationships are applicable only to polygon features.

Delaunay triangulation

In many ways, Delaunay triangulation is an approach for points akin to the contiguity methods for polygons. For point features, a Delaunay triangulation method can be used, which is equivalent to generating Thiessen polygons on the points and using the contiguity edges corners method.

Thiessen polygons divide up the map so that each point in the dataset has a polygon where any location within that polygon would call that point its nearest neighbor. Contiguity is then applied to those polygons to define the neighborhood for each associated point.

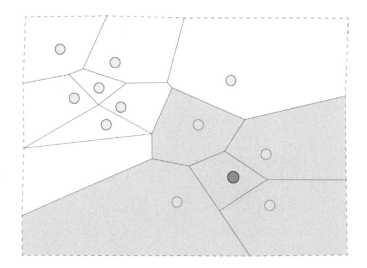

Custom relationships

Of course, the relationships between the features in our data won't always be purely spatial. The world is increasingly interconnected. We can take a plane and be in another hemisphere in a couple of hours, economic transactions are not necessarily constrained by space or time, and the factors that influence us do not always fall within our basic spatial neighborhood.

If none of the purely spatial options are quite what we're looking for, we can choose to create a spatial weights matrix. The spatial weights matrix, while it may have an intimidating name, is actually a very simple file that stores how all features in your dataset relate to all the other features in your dataset. A spatial weights matrix can be manually edited to allow for very customized definitions of spatial relationships that suit our analysis needs and go beyond existing definitions of spatial relationships.

For example, a functional definition of relationships might take additional information into account such as the frequency of interaction between features. New York and Los Angeles are far apart geographically, but if our analysis is related to air travel, it may be important to consider them as part of the same neighborhood. The spatial weights matrix allows for this type of customization.

What's the "right" spatial relationship?

With so many different ways to define what it means to be neighbors, how do we know which one to use? If we were analyzing the cost of procedures at hospitals in an area that served both rural and urban communities, the number of neighbors method would allow for hospitals in rural areas to be related to their closest hospitals, even if those nearby hospitals are very far away by urban standards. If we were using a distance band to define that neighborhood, then the distance required for rural hospitals to have even a single neighbor might be so large that it led to urban hospitals being related to every single hospital in the urban area.

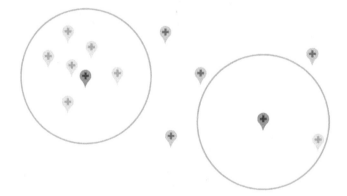

In this case, the number of neighbors method, which adaptively defines neighborhoods, would be a reasonable choice.

Alternatively, if you were analyzing species distribution and you knew that your species of interest traveled only a certain distance from home, a meaningful distance band would allow for a more representative spatial neighborhood. Using a number of neighbors might extend the neighborhood beyond the distance that the species might travel, making distance band a reasonable choice.

Once we've chosen an appropriate conceptualization of spatial relationships, we have to define the characteristics of that neighborhood. If we're using a distance-based approach, should our distance be 500 meters or 3 kilometers? And what impact will that distance have on our analysis? If we're using the number of neighbors, should it be 30 or 300? The truth is, there is no right or wrong answer. But the decision does have a big impact on our analysis. Throughout this book, we'll see examples of algorithms that use spatial relationships in their calculations, and we'll start to get a deeper sense of the impact that this has.

In the meantime, one important concept that comes up here is scale. While there is no right or wrong spatial relationship, the decision that we make does decide the scale of the question that we're asking. Are we looking for clusters in a neighborhood or in a region? Is our question at the local scale, or is it at the national scale? The most important thing to consider is if the choices that we're making are aligned with the question that we're asking and the tangible ways that the results of our analysis will be used. For instance, if we are analyzing access to supermarkets, we might want to use a scale that reflects the distance that people are willing to travel to get to the supermarket.

The bottom line: There is no one-size-fits-all conceptualization of spatial relationships. It's our job to think critically about the scale of our question and choose an approach that is tailored to our specific problem. As analysts and data scientists we are often comforted by concepts that are black and white, but the reality is that there is an art and a science to doing great analysis. We must bring our subject matter expertise to the table and make defensible decisions about what relationship to use, what distance band or number of neighbors is appropriate, and more. And we must be ready to explain the choices we make so that others can feel confident in their merit.

Chapter 2

Means and Medians

Introduction

As we discussed in the previous chapter, spatial statistics help us move from a world of data to a world of information.

A common first step when interpreting raw data is to summarize it using descriptive statistics. These descriptive statistics are a useful way for us to boil down a lot of data into something more tangible and useful. Instead of looking at 100,000 rows of numbers, we might calculate the mean, median, and standard deviation to get a better understanding of our data. These statistics are known as measures of central tendency and dispersion.

With spatial data, we can use these same principles to summarize locations, also known as measuring geographic distributions.

Methods in this chapter can help us answer the following types of questions:

- Where is the center of a disease outbreak?
- What is the area being served by each distribution facility?
- How has the extent of a smoke plume from a fire changed over time?
- How do the home ranges of prey and predator species overlap in a habitat?

Central tendency

Central tendency calculates a single location to summarize a set of locations. It identifies what can be thought of as the "typical" or "average" location, or the one most representative of the entire dataset.

Averages, or other measures of central tendency, are incredibly valuable, but they can also be misleading. This is especially true when our data is not normally distributed. Let's first consider this in the context of data space (meaning we're just looking at a set of numbers, no location).

When data is normally distributed, most of the values fall somewhere in the center of the distribution, with fewer and fewer values above and below that center. In a histogram, a chart that is commonly used to visualize the distribution of a set of numbers, this takes the shape of what is often referred to as a bell-shaped curve.

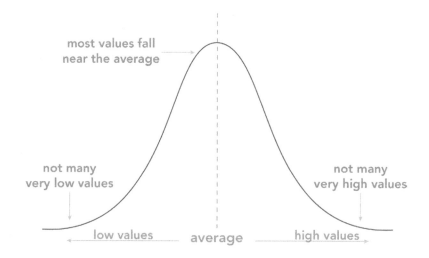

When the data takes this shape, it makes sense that the average is quite representative, because that average falls at the center and that center is where most of the data lies.

What about when data isn't normally distributed? Let's look at an example of what's called a bimodal distribution, where instead of most values falling at the center, we have a histogram that has two peaks.

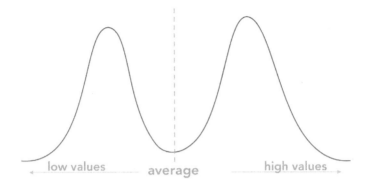

low values   average   high values

These different peaks often represent two different groups mixed together. For instance, in a dataset that combined the number of points scored in all soccer games (usually around 2–3 goals per game) and the number of points scored in all basketball games (usually 105–110 points per game), we'd expect a bimodal distribution because soccer games are usually relatively low scoring and basketball games are usually very high scoring.

We could calculate an average of 50 and, without background knowledge, assume that most games had about 50 points scored, but that wouldn't be useful in understanding either soccer or basketball. It isn't that the average would be wrong, it just wouldn't be useful.

The same can be true if we think about location data. If we have data points throughout the entire United States, an "average" in the middle of the country could very well be representative. But imagine that we have point data in only Los Angeles and New York City, for instance. The average location would be somewhere in the middle of the country, which would not be representative of either New York City or Los Angeles.

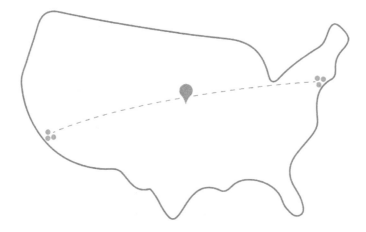

Central tendency may be useful in New York City alone, and it may be useful in Los Angeles alone, but combining the locations into a single dataset and calculating it won't make much sense. For this reason, it is really important to visualize our data to help us understand if and when central tendency will be appropriate, or to determine if the data should be divided into more meaningful groups.

With that said, let's use this set of points to explore the different measures of central tendency.

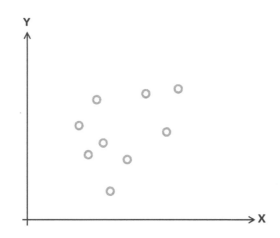

Mean center

Mean center identifies the location on the map that is most central to the features in our dataset.

The calculation for mean center is pretty straightforward. Each point in our dataset has a latitude and longitude value describing its location.

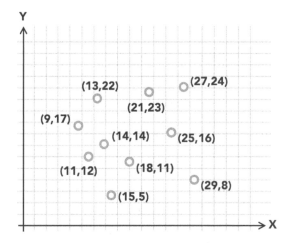

The mean latitude and mean longitude for all the points are calculated, just as we would with any other set of numbers. Then the mean latitude and longitude values are plotted on the map, identifying the mean center.

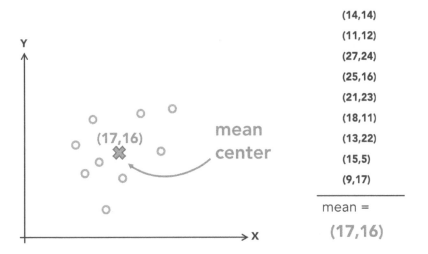

It's as simple as that.

We might use mean center to help us determine where to park a food truck so that it is centrally located to all the businesses in a downtown area. The mean center would be the place that would maximize convenience for all patrons in the area.

Median center

Median center, also known as the center of minimum travel, identifies the location on our map that minimizes the total distance to all features in our dataset. While the mean calculation is as simple as calculating the mean x,y coordinates, the calculation of the median center is computationally a bit more complex. It involves an iterative process to find the approximate location that minimizes the distance to all other features. The algorithm uses the mean center as a starting point and measures the distance to all features. It then iteratively measures the distance to all features from different locations moving away from the mean center until it reaches a threshold where new locations are no longer lowering the total distance from all features. This iterative process is necessary because the set of potential locations to test is infinite, so the median center is by definition an approximation of the location that minimizes total distance to all features.

Just like in traditional statistics, median center is more robust to outliers than mean center and can be used when we do not want our results skewed by a few features on the edges. If we think about the way a traditional median is calculated, by ordering the values and choosing the one in the middle, it makes sense that outliers would have less of an effect than in the traditional calculation of the mean.

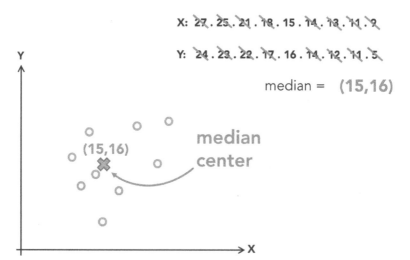

Rather than changing the sum of the values dramatically, an extreme value simply takes its place in line and slightly adjusts the median based on its place in order.

Now, let's add an outlier to our dataset and recalculate. Notice how the mean center shifts much farther than the median center:

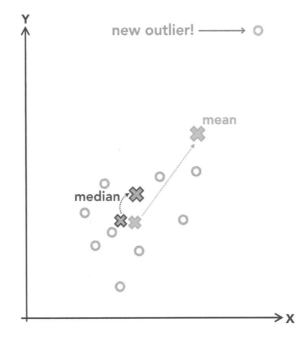

Coming back to our food truck example, what if one of the businesses was far away from the rest? If we use mean center, it will drag the location of our food truck away from the core downtown area and closer to that one spatial outlier. It may even drag us in between the core area and the outlier…putting us close to nobody! If we use median center, though, we will be far less influenced by the outlier. Neither mean nor median is right or wrong, it just depends on how we want outliers to influence the results.

Central feature

Central feature chooses the feature in our dataset that has the shortest total distance to all other features. While median center plots a new location on the map that minimizes that total distance, central feature identifies an existing feature in our dataset that is the most central.

Given our set of points, the distance from every point to every other point is calculated:

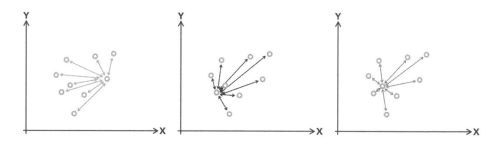

The distances for each feature are summed up, and the one with the shortest total distance is identified as the central feature.

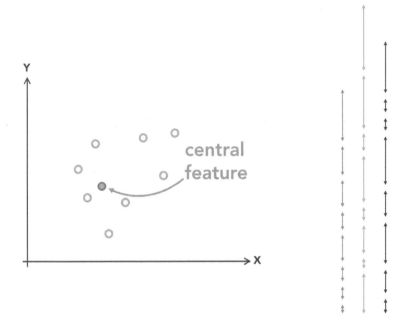

If we were hosting a school-district-wide event for parents, we might use central feature to choose the school that is most convenient for all the parents in the district. In this case central feature is appropriate because we need to host the event at one of the existing schools in our dataset.

Linear directional mean

Linear directional mean calculates the average angle for a set of lines, measured from a common tangent (which you can think of as a horizontal line from which the other angles are being measured).

Linear directional mean can be calculated in one of two ways: direction or orientation. When calculating direction, each line has a start and an end point that are used to calculate the angle. When calculating orientation, no start and end point are used.

Here, we can see two lines that have opposite directions based on their start and end points, but the exact same orientation.

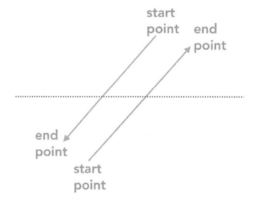

When calculating orientation, these lines have the same angle, and all angles will be between 0° and 180°.

When calculating direction, angles can span from 0° to 360°.

The output of linear directional mean is a single line with a length equal to the average length of all the input lines, and an orientation or direction reflective of the average of all orientations or directions of the input lines.

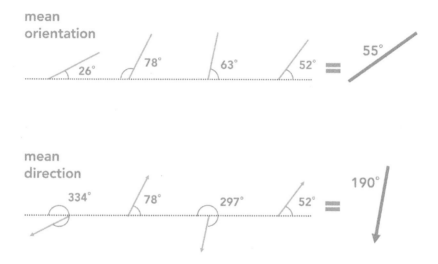

We can use linear directional mean to understand patterns in the direction of hurricane tracks. If we have hundreds of hurricane tracks over several years, we might want to get a sense of the average length and direction. Or we might want to run it for each year to see how average length and direction change over time.

Dispersion

Dispersion describes how spread out or compact a dataset is around its location of central tendency. One common way to achieve this is by measuring how far from the mean center the data points fall. Measures of dispersion give us a sense of the spatial distribution of our data.

This builds on concepts from statistics around the behavior of a normal distribution. Remember, when data is normally distributed, most of the values fall somewhere in the center of the distribution, with fewer and fewer values above and below that center. With a normal distribution of values (ignoring space for just a minute), we expect that 68% of our data will fall within 1 standard deviation of the mean. We expect 95% to fall within 2 standard deviations. And 99.7% to fall within 3 standard deviations.

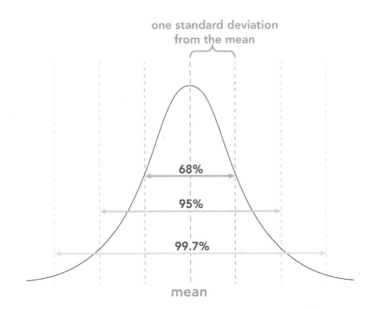

Now imagine that histogram in map form. A spatial normal distribution will be most dense in the middle, near the mean center, and become more and more sparse as we move outward.

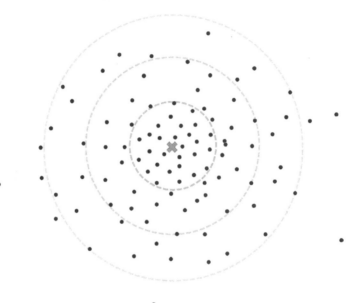

Under these conditions, we expect to see about 63% of the data fall within 1 standard deviation, 98% within 2 standard deviations, and 99.9% within 3 standard deviations. These values are slightly different from what we expect in a normal distribution of values because of the special characteristics of spatial data (namely that there is more than one dimension), but the concepts are quite similar and useful for understanding conceptually how this works.

Let's use the same set of points as an example to explore the different methods available for measuring the spread of our features. Both methods begin by identifying the mean center of the dataset and choosing the features to include in the analysis based on 1, 2 or 3 standard deviations from that mean center.

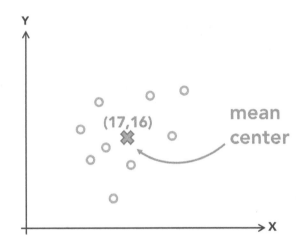

Standard distance

Standard distance is calculated by finding the average distance of each feature to the mean center. That distance is then used as the radius of a circle centered on the mean center. This circle representing the standard distance provides information about the dispersion or compactness of the data. The standard distance is the spatial equivalent of a standard deviation, which measures the distance in data space of each value to the mean. Just like a standard deviation, the standard distance is a useful measure for understanding how representative the mean center is. The standard distance is also a useful way to see where on the map most of our data is concentrated.

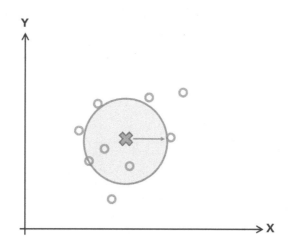

Standard distance could be used to measure the distribution of a tree species in a conservation area or to see how that distribution overlaps with the distribution of bark beetles that are wiping out forests in the area.

Standard deviational ellipse

In addition to measuring dispersion, standard deviational ellipses also measure orientation. In this way, the standard deviational ellipse gives us valuable information that we don't get from standard distance.

To calculate a standard deviational ellipse, the distance from each point's x-coordinate to the mean center's x-coordinate is calculated, as is the distance from each y-coordinate to the mean center's y-coordinate. The average x distance and average y distance become the lengths of the axes of the ellipse. Then, a rotation is calculated that represents the orientation of the underlying data. Ultimately, the ellipse minimizes the distance to all features from the axes, making it a useful way to represent both orientation and dispersion of the data.

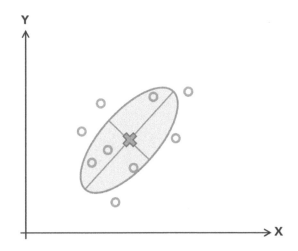

Going back to our bark beetle example from standard distance, we could use standard deviational ellipses to measure not only the distribution of the bark beetles but also the direction of their spread. This could be useful in understanding where we might expect to see them in the future so we can try to mitigate their impact.

Additional dimensions

Sometimes our data has more than just x,y coordinates. Maybe we have a location and an associated attribute, like school locations and the associated number of students enrolled. Or maybe our data is 3D, and has x,y, and z-coordinates. These additional dimensions can optionally be included in the calculations.

Weights

With all measures of central tendency and dispersion, features can be weighted by an attribute to give them more influence on the calculation. In this way, features with larger values for a specified attribute pull the average towards them, exerting more influence on the resulting measure of central tendency or dispersion.

For example, with our school locations, mean center can be weighted by the number of students at each school. By weighting the number of students at each school, the mean center finds the central location of total students, not the central location of schools. Depending on our analysis question, one or the other may be more appropriate.

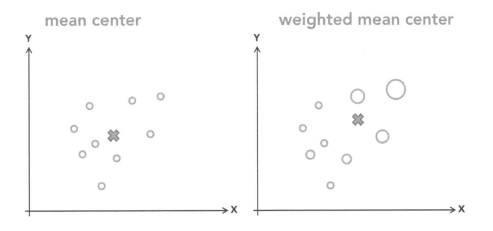

3D

Each of these methods can also be calculated for 3D data using its z-value. Conceptually, the z-value is treated as another coordinate, much like x and y, where the mean x-, mean y-, and mean z-values are all used in the calculation.

For example, mean center can be calculated for centrality in a multilevel building based on office occupancy, or an ellipse can be calculated to summarize species observations underwater or in the air.

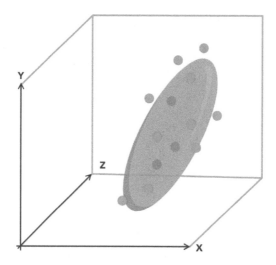

Comparisons

Now we understand how these methods work, but what makes them useful? The reality is, an average is most useful when we have something to compare it to. For instance, knowing the average home value for a town that we're considering moving to is useful, but knowing how that average compares to neighboring towns is even more useful. Or knowing how that average compares to previous years. Ultimately, these comparisons play a key role in how we use that statistic to make a decision.

We often think about two key types of useful comparisons for spatial data. One is comparing the centrality or distribution of different datasets, or different categories within a single dataset. For instance, the standard deviational ellipse for lion tracks compared to the ellipse for elephant tracks in a particular study area. Do they overlap? Are they separated by natural features? Is one more dispersed or compact than the other?

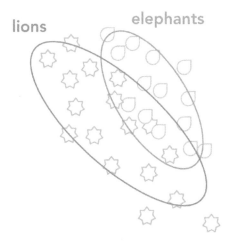

Another useful comparison is across time. For example, we can compare the median center of customer locations from year to year. Is the median center moving? Have our marketing efforts expanded our customer base and helped us move into new areas?

median center of customer locations

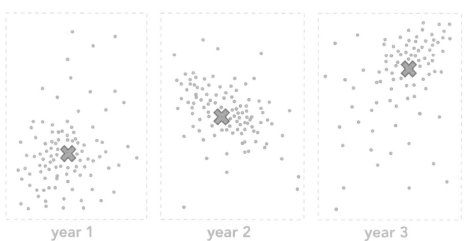

year 1   year 2   year 3

All our examples have included points, but there are actually some interesting applications of these methods for polygons as well. It may seem counterintuitive, since often polygon locations are somewhat fixed (think census boundaries), but when we include a weight in the analysis, we are able to, once again, compare how patterns change over time. For instance, we can weight polygons by population and calculate the mean center over time, allowing us to see how the population has shifted.

population-weighted mean center

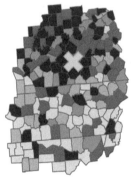 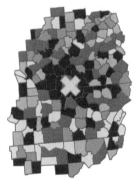 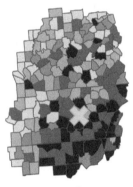

year 1   year 2   year 3

Keep it simple

These methods are very simple and straightforward, but that doesn't mean that they aren't valuable in the analysis process. We often use these measures of central tendency and distribution to get to know our data. Many times we use these tools as part of formulating our hypothesis. Much like with traditional statistics, we don't calculate an average and pat ourselves on the back and call it a day. Instead, we use these exploratory techniques to better understand our data, and in many ways understanding our data helps us determine what questions we can ask and how we want to move forward.

Another point worth mentioning is that the simplicity of these methods is actually one of their greatest strengths. There is a concept in science called Occam's razor that prioritizes choosing the simplest feasible approach when solving a problem. Put another way, reportedly by Albert Einstein himself: "Everything should be made as simple as possible, but no simpler." So, if we can get to the answer in multiple ways, the simplest way is most likely the right way. And we have the added bonus of it being interpretable and approachable for decision makers and others. Just another reason to fall in love with means and medians.

Chapter 3

Finding clusters with machine learning

Introduction

We've already mentioned the idea that our brains are these really powerful statistical machines, and finding clusters is one of the things that we're particularly good at. We use clusters to interpret the world around us. It is an important part of who we are as people and how we navigate our world. And this applies to the way we interpret maps and spatial data as well. When we look at a map, we naturally start to find clusters and make sense of patterns. We can use spatial statistics to go beyond a visual cluster analysis to help us quantify the patterns that we're seeing and also to automate the process.

There are many different techniques for finding clusters. Two high-level approaches that we will cover in this book are machine learning–based approaches and statistical approaches.

Machine learning–based approaches to clustering are increasingly popular and widely applicable to a broad range of problems. They are generally unsupervised techniques, meaning that we don't tell the algorithm ahead of time what it means to be a cluster—it just looks at our data and figures it out for us. In other words, the algorithms detect clusters in our data through evaluating the underlying structure or characteristics of the data (density, distances between features and attribute values, etc.).

The other high-level approach is the more statistical approach to clustering where we're asking a different kind of question. Ultimately with statistical clustering methods, we want to know if a spatial pattern we're seeing is meaningful, so we ask the question: What are the chances that this spatial pattern in our data occurred due to random chance? Is this spatial distribution of values random? Or is there a pattern?

We know it can be a bit overwhelming when there are so many techniques out there that fit into the bucket of "clustering." Whether we're looking for areas with unexpectedly high values, or finding groups of features with similar characteristics, or finding point patterns in the landscape, these approaches can all be referred to as "cluster analysis." Ultimately, each method asks a different question. There isn't a right or wrong method to use, just the one that most appropriately answers our question.

In this chapter, we will explore how to identify inherent clusters in our data based on similarities in location and/or attribute values using machine learning–based clustering methods. In the next chapter, you will learn about statistical clusters, and how we can evaluate them to find nonrandom patterns in our data.

Methods in this chapter can help us answer the following types of questions:

- Which intersections have clusters of fatal car accidents?

- How can I partition sales regions so that they have even populations and equal areas?

- What are the areas with similar vulnerability characteristics based on socioeconomic status, governance, population density, and climate change?

Density-based clustering

Density-based clustering is a machine learning–based approach that finds clusters of points based purely on their location, finding groups of points that are clumped or grouped close together. We'll be discussing three different methods for identifying density-based clusters: DBSCAN (defined distance), HDBSCAN (self-adjusting), and OPTICS (multiscale). While the algorithms for each approach differ, they all detect clusters that reflect the underlying densities in the data.

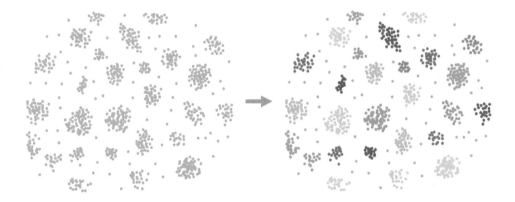

So, let's get into how these methods work. Despite their differences, all three methods require us to define the minimum number of features necessary for a group of points to be considered a cluster. This is one of the places that we define our question and use our subject matter expertise to guide the algorithm in the right direction. For instance, if we were looking for clusters of traffic accidents, we might specify that we need at least 10 accidents in close proximity to be considered a cluster.

Based on this value, every feature has a core distance. A core distance is the distance from each feature at which the minimum number of features required per cluster is reached.

For example, the minimum number of features required per cluster in this illustration is five (and note that the focal feature is included in that count):

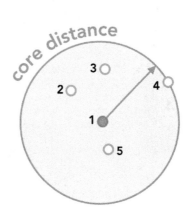

This concept of a core distance is shared among the algorithms, but this is where they diverge. Next, we'll explore how each method makes use of the number of features and the corresponding core distance to find clusters in our data.

DBSCAN–defined distance

DBSCAN looks for clusters of similar densities by going point by point through the dataset and evaluating each point based on a comparison of its core distance, which we just discussed, and a user-defined search distance. This search distance represents the spatial scale at which we are searching for clusters. So, for a given point we calculate the core distance based on the number of neighbors specified. We then compare that core distance to the search distance to determine if that point is part of a cluster. When the core distance is smaller than the search distance, the algorithm is able to find the specified number of points within the specified search distance, and that point is marked as part of a cluster. When the core distance is greater than the search distance, the algorithm has to go out beyond the search distance to find enough neighbors to constitute a cluster, which means that point is marked as noise.

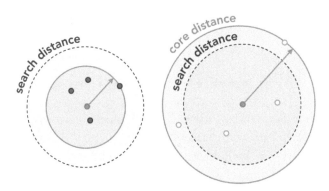

It is for this reason that we refer to DBSCAN as a defined-distance approach. It employs a fixed search distance across the entire study area. This results in clusters that have similar densities because of the rigid search distance that's imposed. If we have a clear search distance, it's a good method to use.

HDBSCAN—self-adjusting

HDBSCAN uses hierarchical clustering to look for clusters of varying densities. It is a data-driven approach, so it finds hierarchical clusters in the data, which you can think of as clusters within clusters. This is often illustrated using a dendrogram, which shows the hierarchy that is used to distinguish meaningful clusters.

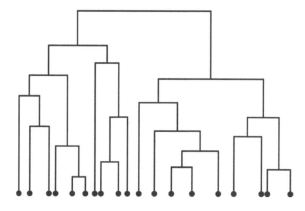

Conceptually, the algorithm begins by creating a hierarchy that combines the smallest clusters into larger and larger groups of clusters. It then evaluates the hierarchy to figure out which clusters, or groups of clusters, are distinct or "stable" enough to be considered an output cluster. Sometimes that will be a single cluster that meets the minimum number of features requirement (at the bottom of the hierarchy), sometimes it will be a grouping of clusters together. This type of hierarchical clustering allows us to decide what becomes a cluster using more than just a cutoff distance (which inherently imposes a fixed density on all output clusters); instead, it looks at the stability at each level of the hierarchy to choose which points belong in the same cluster.

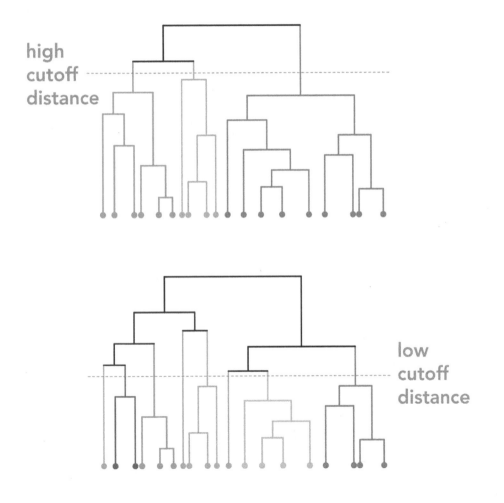

cutoff
distance

low
cutoff
distance

It is for this reason that we refer to HDBSCAN as a self-adjusting approach. Under the hood, HDBSCAN is actually pretty algorithmically complex and uses a number of machine learning techniques to determine the appropriate cutoffs. Luckily, conceptually it is quite simple to understand because this concept of fixed vs. adaptive spatial relationships is a common thread throughout spatial statistics and the way we think spatially about our data.

OPTICS—multiscale

The last method for density-based clustering is called OPTICS. OPTICS works by going from feature to feature and measuring the distance from each feature to the next unvisited closest feature and creating what we call a reachability plot.

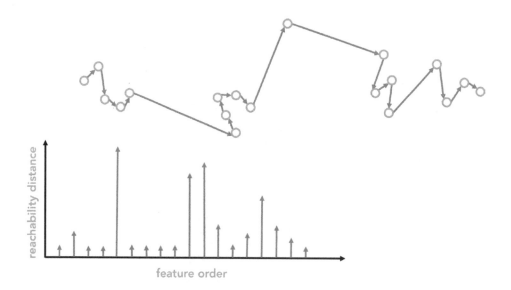

The peaks and valleys in the reachability plot are then used to define the clusters. Short distances become valleys and long distances become peaks. When we have short distances, it means that a series of features are near each other, and those become clusters. When we have long distances, it represents either a separation between clusters or points that will be marked as noise.

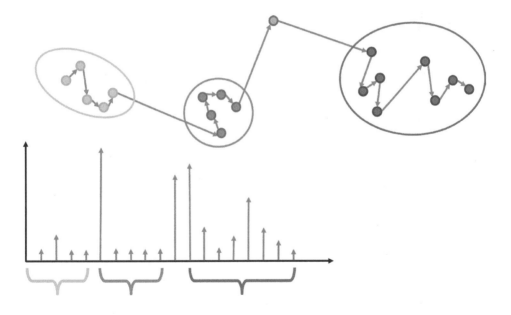

OPTICS is unique in that it has the most flexibility when fine-tuning parameters. A parameter called cluster sensitivity determines the thresholds used to define what's considered a peak and what's considered a valley. After all, one person's peak is another person's valley. In this example, with a low cluster sensitivity we might have three clusters, where the blue group, despite having what some might call a peak in the middle of it, is considered one cluster.

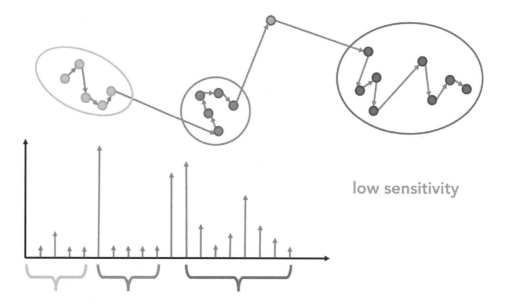

low sensitivity

A higher cluster sensitivity lowers the threshold for how high a peak needs to be relative to its valley to separate clusters, and we end up with four clusters instead of three, as we see in this example.

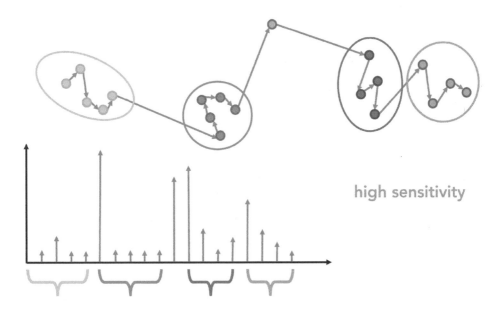

high sensitivity

It is important to note that the way the cluster sensitivity parameter chooses what is considered a peak is relative to that peak's valley. So, we aren't saying that a particular overall distance is required for something to be a peak, but rather a peak is determined by how big that distance is compared to the distances in its valley. This allows OPTICS to find clusters of variable densities, which is an important characteristic of both OPTICS and HDBSCAN.

This ability to adjust the cluster sensitivity and choose how we want our clusters to be defined is a particularly powerful aspect of the flexible OPTICS algorithm.

What's the "right" density-based clustering algorithm?

Let's summarize these different methods. DBSCAN uses a fixed search distance. It finds clusters of similar densities and it's pretty fast. HDBSCAN uses a range of distances to find clusters of varying densities and it's very data-driven and requires the least user input. Note that HDBSCAN also has very little flexibility, so we have little control over the clusters that it finds. OPTICS uses neighbor distances to create a reachability plot. OPTICS is the most flexible for fine-tuning but can also be the most computationally intensive.

So, how do we choose the "right" one? The truth is, when doing this kind of cluster detection, we often have a notion of the types of clusters that we want the algorithm to find. In fact, we might even be able to circle these clusters on the map by hand. But the power of machine learning is that we can automate this detection, so that we can save time and resources. The "right" method, in this case, is often the one that does the best job finding the clusters that we would have picked out ourselves. The one that seems to reflect our notion of clusters best. And the way to figure that out is often to give them each a try. We know this likely isn't the answer you were hoping for, but analysis is inherently iterative, and the best results come when we've explored our options thoroughly.

The four (eight) color theorem

Density-based clustering often finds many more clusters in our data than our eyes can distinguish using unique colors. Since rendering each cluster with a unique color is unrealistic, we have to get creative. One approach that has been used for years in the creation of atlases is something called the four color theorem. Trying to make every country a unique color would be impossible, but making sure that adjacent countries aren't symbolized using the same color is achievable. That's where the four-color theorem comes in, and we use a variation of this technique to visualize the results of our density-based clusters. Using eight unique colors, the method ensures that no two adjacent clusters will be assigned the same color. So, while there may be many light-green clusters in the map, they should always have clear boundaries.

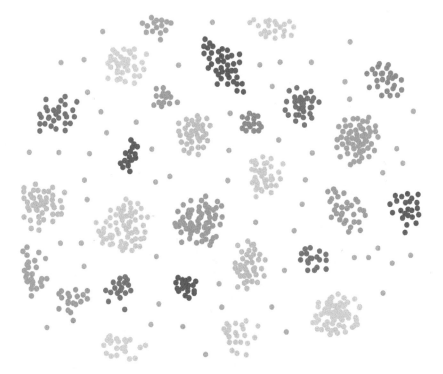

Multivariate clustering

So far, we've looked at clustering based purely on location. Now we're going to look at clustering based on feature attributes or values. Multivariate clustering is an approach to classification. As we mentioned, organizing things into classes or categories is a quintessential human trait essential to our survival. Classes are useful abstractions about a group of features that have similar characteristics. So classification serves as a type of heuristic, or shortcut, that allows us to learn or infer characteristics about a feature or entity based on what class it belongs to. For example, we might infer that a car will have a third row of seats because it is classified as a minivan.

The goal with these methods is to create distinct clusters of features, where attribute values are similar within clusters and dissimilar between clusters. In other words, we want features within a group to be really alike, and we want each group to be very different from the others.

Whereas with density-based clustering, some features were found to belong to clusters and some were marked as noise, with multivariate clustering, all features will be partitioned into a cluster.

One of the most popular ways to do this type of multivariate clustering is using a method called k-means. K-means is not a spatial algorithm; instead, it looks at the values in data space. Understanding this data space is useful when thinking about how the algorithm works.

Let's say we have a dataset with three variables: population density (PopDensity), average income (AvIncome), and median age (MedianAge).

PopDensity	AvIncome	MedianAge

"Data space" refers to where each feature falls in a plot based on its attribute values.

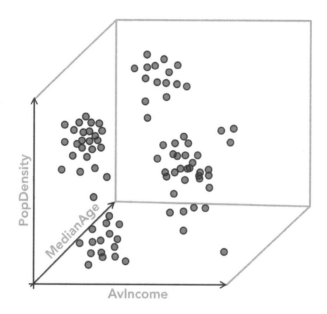

Within this data space, we can see that features cluster together based on their values. K-means will partition features based on any number of numeric attributes (only three are pictured here because…good luck illustrating more than three dimensions) into however many clusters we specify. The k in k-means refers to the number of clusters that we choose.

Features are split into clusters based on where they fall in data space and how many clusters are specified. Features that are close together in data space are lumped together into a cluster.

Here we can see what our clusters might look like if we asked for two, three, four, and five clusters to be found in our dataset:

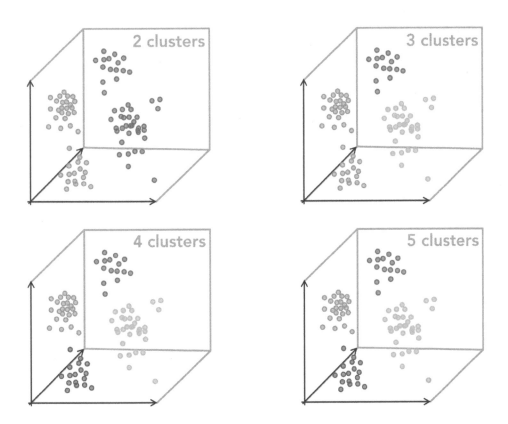

Conceptually, k-means picks a random set of seeds, or starting points, based on our specified number of clusters and iteratively finds ideal cluster centers that are as different as possible from each other. It then assigns each feature to one of those centers so that the distance to those centers (in data space) is minimized among the groups. Because we are trying to minimize distances to those centers, we can also use those distances to quantify how effective our clustering has been. This can be useful when determining an ideal number of groups based on the natural clustering in the data, and also when trying to determine if one set of variables is more useful than another for classification.

Choosing the most useful number of clusters (k) often requires trial and error, and a bit of patience. One reason is that the result is very sensitive to the k that is chosen. If our data has 5 natural clusters in it, but we choose a k of 3, we will fail to separate clusters that would have been meaningful. Alternatively, if we choose a k of 7, we will separate meaningful clusters unnecessarily. A number of methods are available to evaluate the distinctness of clusters, including a popular approach called the Calinski-Harabasz pseudo-F statistic. The pseudo-F statistic quantifies the quality or distinctness of clusters for increasing numbers of clusters using a ratio that reflects within-group similarity and between-group difference. This statistic is calculated for each potential number of clusters.

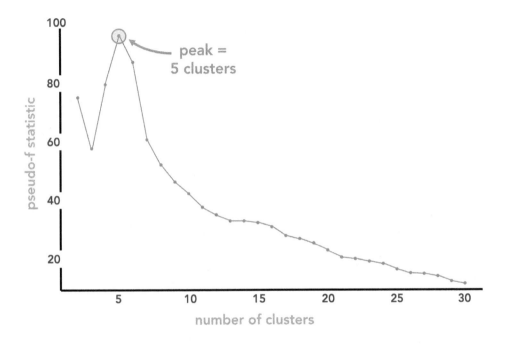

Charting the results, we look for the peak pseudo-F statistic because it represents the number of clusters that will lead to the most distinct results.

You might be wondering what happens when our attributes have different units, which often results in values of dramatically different magnitudes (for instance, average income can be in the tens of thousands, and age has a pretty strong cutoff somewhere below 200). To get around this, one common approach is the standardization of the values being analyzed. Standardization maintains the shape of the distribution of each attribute relative to its mean using something called a z-transform. The math is as simple as taking each feature's value, subtracting the mean, and dividing it by the standard deviation, which centers each variable's mean at zero. This standardization ensures that all variables contribute appropriately to the result, regardless of their magnitude.

PopDensity   **124** people/sq mile   ⟶   0.792

AvIncome   **73,847** dollars   ⟶   -2.761

MedianAge   **43** years   ⟶   1.225

It essentially puts all the values on the same scale. This ensures that average income doesn't take over the clustering and result in clusters that are only representative of differences in average income.

While k-means is not inherently a spatial method (remember, it uses just attribute values, and no concept of spatial relationships), when we map the results, we usually see spatial patterns worth exploring. This makes sense, because we expect things that are closer together to be more related than things that are farther apart.

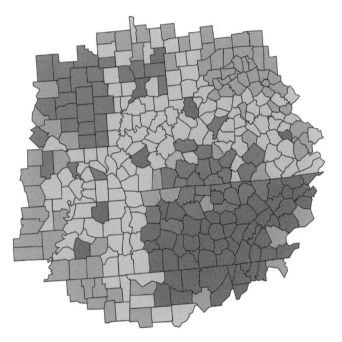

So, while the method isn't spatial, a spatial approach to interpreting results adds a lot of value. If we ignore those spatial patterns, we're leaving a lot of information on the table.

Because this multivariate clustering is all about attribute values, a map of the distinct groups will never be enough for us to interpret the results. We also need a way to understand the characteristics of each cluster. We can use box plots to do just that.

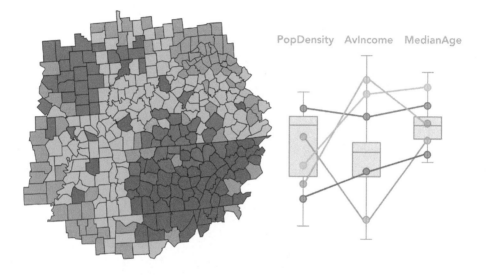

Here we can see that the box plots help us interpret the characteristics of our clusters. The colors in the box plot correspond to the colors in the map. The gray box plot represents the overall distribution for the entire dataset. The colored points represent the values for each individual cluster. The green cluster, for instance, has roughly average population density but is well above average in terms of income and median age. Alternatively, the purple cluster is well below average for population density and median age, but about average in terms of income. These box plots, in combination with the map, give us a sense of what each group really represents.

Spatially constrained multivariate clustering

With multivariate clustering, our goal is to create groups of features that have similar characteristics. But sometimes we need those groups to be contiguous, which introduces an additional constraint to our clustering. For instance, we might want to delineate areas for a targeted marketing campaign, so that we can tailor our messaging based on the local characteristics of each region. This is where spatially constrained multivariate clustering comes in (we know, it's a long name, but at least it's very descriptive).

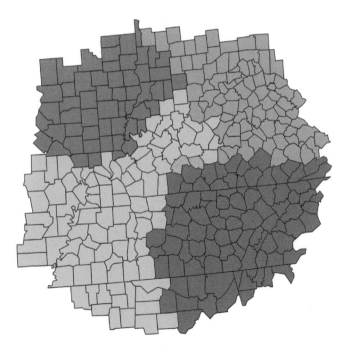

Like multivariate clustering, spatially constrained multivariate clustering also partitions features into clusters based on their similarity in attribute values, but with this method we also use spatial constraints to ensure that the resulting clusters are spatially contiguous. In this way, the approach uses both attribute values *and* location to create clusters.

The spatially constrained multivariate clustering approach uses a connectivity graph (also known as a minimum spanning tree) to lay out the features in data space and "spatial space." The set of features is linked based on how far they are from each other based on location and attribute values.

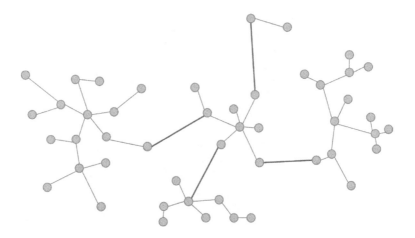

Those links are then broken in ways that keep the clusters as distinct as possible.

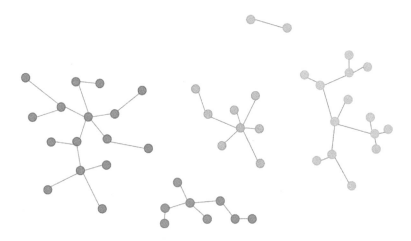

It is important to note that when we impose spatial constraints, clusters will almost always be less distinct than they would have been with no spatial constraints. For instance, adding that spatial constraint might lead to contiguous neighbors that don't have much in common being forced into the same cluster just to satisfy the contiguity constraints. Because of this, some features have stronger similarities to their cluster than others. This can be measured by calculating membership probabilities.

A feature's membership probability tells us how strongly allied the feature is to its cluster. Features with strong membership probabilities are exemplar features; their attribute values are most representative of their cluster. Conversely, features with weak membership probabilities are more wishy-washy, and they could be lumped into a neighboring cluster without too much impact on overall cluster distinctiveness.

Just like with multivariate clustering, the map output shows us which features belong in each cluster, but to understand the distinctive characteristics of each cluster, we can examine the box plots.

Being able to do this kind of spatially constrained multivariate clustering is very valuable when our goal is to create contiguous regions with attribute values that are as similar as possible.

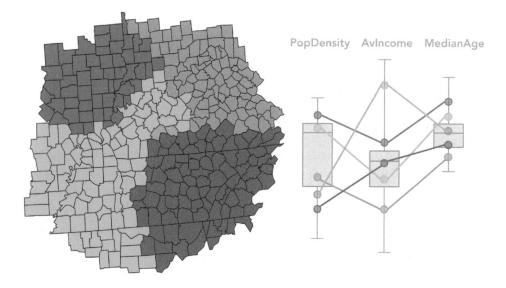

PopDensity   AvIncome   MedianAge

Build balanced zones

While the multivariate clustering methods group our data into clusters that are different from each other, sometimes what we're after is clusters that are alike. Conceptually, build balanced zones does the opposite of multivariate clustering. It partitions our data into groups so that the groups are as uniform or as balanced as possible. This can be as simple as making sure each group has the same number of features or a similar sum of a variable of interest. For instance, we might use build balanced zones to design sales territories so that they are balanced in terms of population and other characteristics or to delineate zones for canvassing that balance the area as well as the number of homes within each zone.

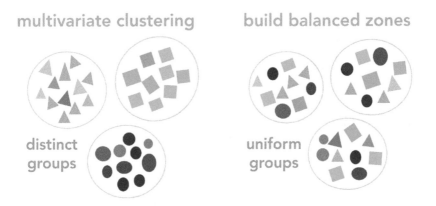

multivariate clustering   build balanced zones

distinct groups   uniform groups

This is accomplished using a really cool optimization method called a genetic growth algorithm. Let's say we wanted to create four zones. The algorithm starts by choosing one of the features as a random seed and lumping nearby features together until the zone criteria are met. Then another random seed is chosen, neighbors are lumped in until the criteria are met, and so on until we have four zones based on four random seeds.

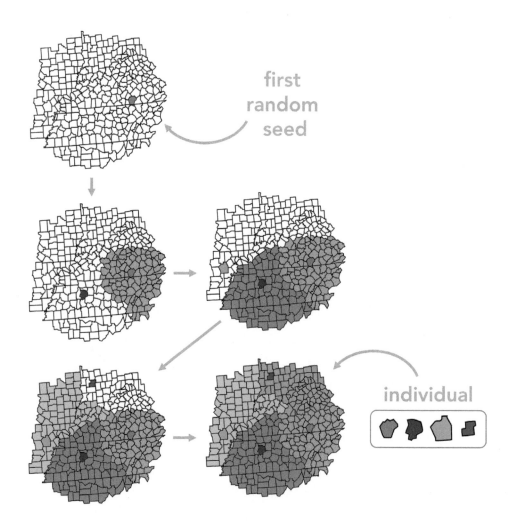

first
random
seed

individual

This set of seeds makes up one potential solution. Each potential solution is called an individual. The genetic algorithm gets its name because an entire population of zone individuals will compete against each other in a "survival of the fittest" manner until the optimal solution is chosen. Each individual receives a fitness score. Individuals with lower fitness scores are considered the most fit.

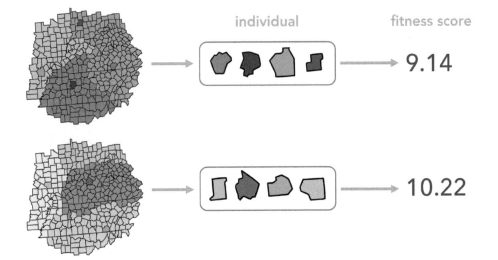

individual   fitness score

9.14

10.22

Only the fittest individuals, the top 50%, move forward to create offspring in the next generation. These best-scoring individuals then metaphorically mate with each other to create new sets of seeds and clusters.

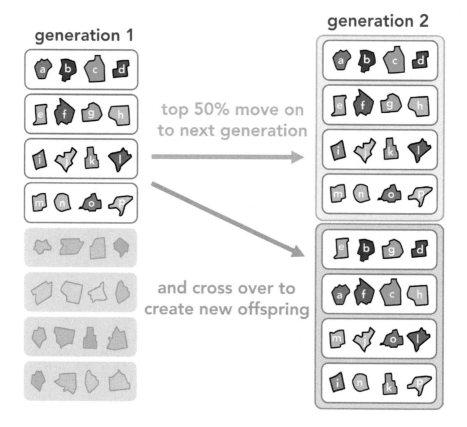

This competition goes on for a specified number of generations. The algorithm also makes use of other concepts from genetics like random mutations to increase diversity and approach an optimal solution.

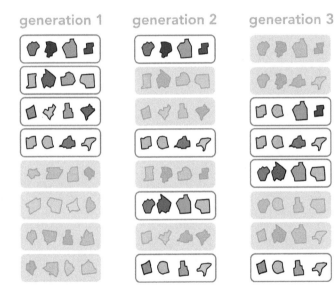

Ideally, we iterate through these generations until we are no longer improving the scores of our zones. We can use a chart like the one below to get a sense of the point of diminishing returns.

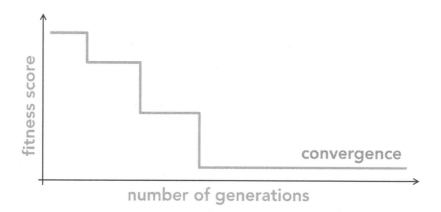

The genetic algorithm is particularly useful here because exhaustive evaluation of all possible combinations is not possible, so a heuristic is our best bet for finding an optimal solution.

Through this natural selection, we can approach an optimal solution for a combinatorically complex problem that has a daunting number of potential solutions. Pretty amazing.

Summary

Each of these methods allows us to find different types of clusters and answer very different questions. With density-based clustering, we can find spatial clusters in our point data. With multivariate clustering and spatially constrained multivariate clustering, we can bring attributes into the mix and create partitions that have similar characteristics within groups but are different between groups. And build balanced zones allows us to do the opposite, creating partitions that are as similar to each other as possible.

We hope that walking through each of these machine learning algorithms and how they work conceptually helps make these techniques a bit more tangible and approachable. They may not be rocket science, but it is pretty amazing to see the way that these algorithms build on the way our brains work to automate the detection and creation of these diverse types of clusters.

Chapter 4

Statistical cluster analysis

Introduction

In the last chapter, we learned about detecting natural groupings in our data using machine learning–based clustering methods.

In this chapter, we're going to explore clusters of a different variety—statistical clusters.

The statistical clustering methods we'll cover in this chapter are used to measure local clustering of an attribute's values over space. We're looking for places where high values cluster together and low values cluster together.

Methods in this chapter can help us answer the following types of questions:

- Where are there clusters of low test scores for K-12 students?

- Where are the hospitals with higher readmission rates than their neighbors?

- Where are there clusters of citizens without internet access?

Subjectivity of maps

This is a chapter about statistical cluster analysis, but before we dig into just what that entails, let's take a quick detour to explore a concept very near and dear to our hearts, the subjectivity of maps.

Earlier, we discussed how we can turn our data into information using spatial statistics. We proposed that merely plotting points on a map is rarely (if ever) enough to give us a deep understanding of our data.

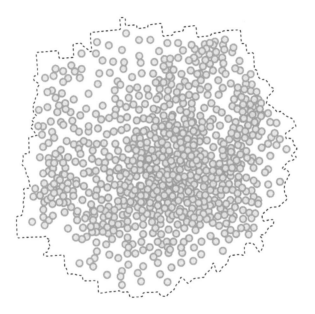

One of the biggest challenges with points on a map is that patterns can be hidden by overlapping and coincident points. In very dense areas, it can be difficult to tell which places are, in fact, more dense than others.

A common next step when mapping point data is to create a heat map, which shows where we have high and low concentrations of points across the study area.

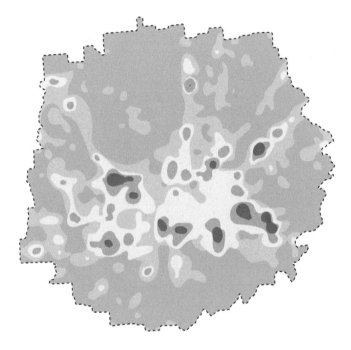

These visualizations can be a great way to start exploring our data and get around some of the challenges introduced by simple point maps. But heat maps introduce their own challenges. Take these two heat maps, for example.

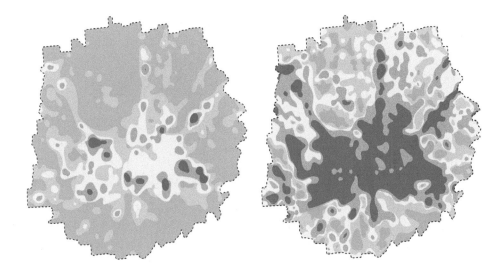

The map on the left looks pretty sparse, not a lot going on. The map on the right looks a lot more dense, a lot more intense. Each map tells a very different story.

But these maps represent the exact same data. The only difference is that we've made different decisions about what it means to be red and what it means to be blue. In the map on the left, we used a natural breaks classification method. In the one on the right, we used a quantile classification method.

natural breaks   quantile

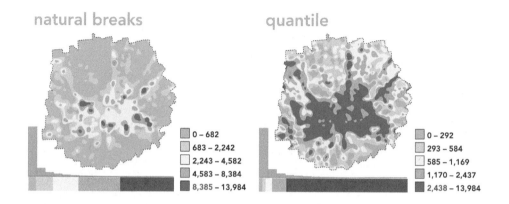

So, on the right, all that was required was a value of **2,438** to be considered red. In the one on the left, a value of **8,385** was required. Same data, different classification…very different story.

And the same is true with a thematic map. Change the classification method, change the story. In some cases, pretty dramatically.

natural breaks   quantile

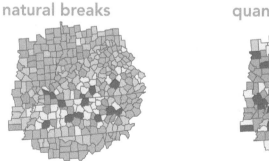
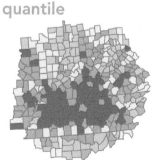

Most people who have never made a map (which is most people) think about maps as objective truths. If we choose the map on the right, they believe us. If we choose the map on the left, they believe us. And both are "correct." They're both telling the truth. But they are also both telling a story. So, we have a lot of power when we make these maps, and with great power comes great responsibility (as Spiderman's uncle famously said).

While maps can be incredibly powerful communication tools, we do have to be careful. We have the power to tell different stories with the same data. Heat maps and thematic maps are useful and are absolutely a great place to start…but we can do more!

And one of the ways that we can do more and reduce this subjectivity is using statistical clustering techniques like hot spot analysis. This is an example of using a hot spot analysis on that same point data. It may appear at first glance that this map looks just as subjective as the others since there are still blues and reds. Under the hood, though, what's being used to determine what's red and what's blue is statistics.

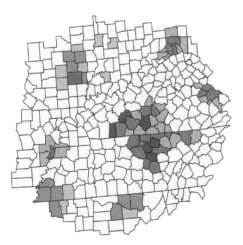

Here, the areas in red are statistically significant hot spots, places where high values are clustered together. The areas in blue are statistically significant cold spots, where low values are clustered together.

So what do we mean when we say "statistically significant"? To talk about that, we'll have to talk a bit about inferential statistics.

Probabilities, confidence, and randomness

Inferential statistics are about testing hypotheses.

In the case of spatial statistics, we're often testing a null hypothesis referred to as complete spatial randomness. Complete spatial randomness is what we would expect to see if there were no pattern, no spatial process—a completely random spatial distribution.

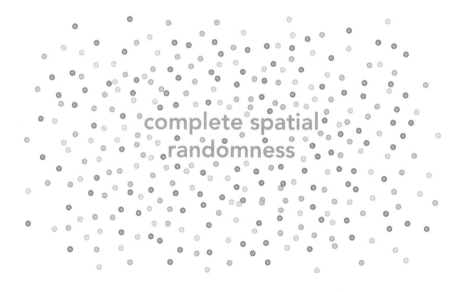

When a pattern is not random, it indicates that there are underlying spatial processes that are worth exploring and that we might even be able to impact or remediate.

So, statistical clustering techniques use inferential statistics to ask the questions, "What are the chances that this pattern that we're seeing happened randomly? How likely is it that this pattern is the result of a random spatial process?"

To illustrate this concept, let's take a look at a hypothetical set of features, each with an associated value:

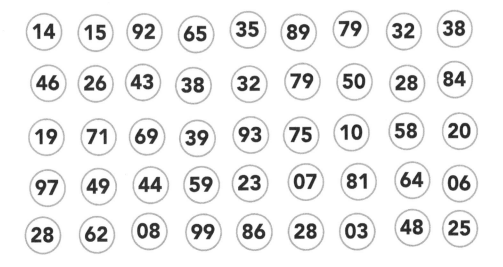

If we were to pick up our features, scramble them, and drop them back down, we'd expect high and low values to end up randomly scattered throughout…a little here, a little there.

But when extreme values all end up clustered together, that is different from what we'd expect to see under typical random circumstances.

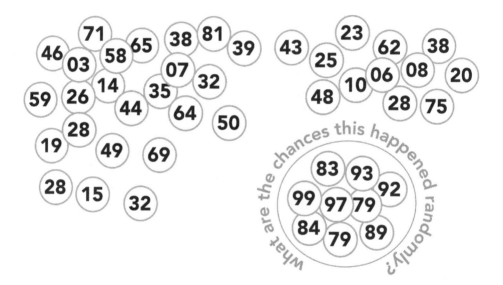

Do you think it's likely that all the high values just happened to land near each other randomly?

Neither do we.

But it's not always this obvious.

And we answer that question using p-values and z-scores. Because we're asking the question, "How likely is it that this pattern is random?" we start by seeing how far it is from random using z-scores. Z-scores are essentially standard deviations from the mean: How far is the value from what we would expect if it was random?

The larger the z-score, the less likely it is that a random process would produce the data, and this likelihood is represented using p-values. The p-value is the probability of a completely random process producing the amount of clustering seen in the data. You can think about the p-value as the evidence against the null hypothesis. So, a p-value of 0.01 tells us that there is a 1% chance that data generated from a random process would have the amount of clustering seen in the data. Since this probability is low, we can conclude that the pattern is not random with 99% confidence. Which means we can be really confident that we're looking at a meaningful cluster (we've got 99% confidence, in fact).

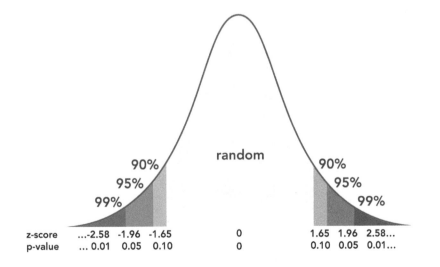

	random	
90%		90%
95%		95%
99%		99%

z-score	...-2.58 -1.96 -1.65	0	1.65 1.96 2.58...
p-value	... 0.01 0.05 0.10	0	0.10 0.05 0.01...

A feature with a high z-score and a low p-value is very unlikely generated by a random spatial distribution. In other words, we can be confident that it belongs to a nonrandom cluster. These are commonly referred to as statistically significant clusters.

Features belonging to nonrandom clusters are classified into three ranges: 90%, 95%, or 99% confidence at each extreme. So, instead of choosing a pretty arbitrary classification method, the output map symbology corresponds to these levels of confidence. And when we know that we have significant clustering, we can be confident that we aren't focusing our attention on random clusters…and are instead focusing on areas that represent real, underlying spatial processes.

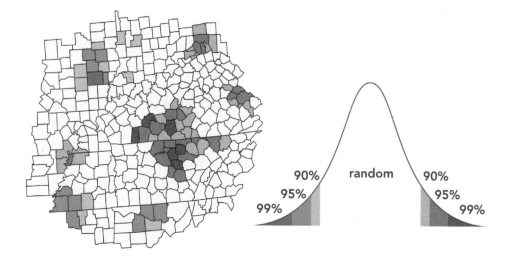

It is important to note that while statistical clustering techniques can help us minimize the subjectivity in our pattern analysis, they do not eliminate all subjectivity. All analysis involves some amount of subjectivity. We choose what data to analyze, what questions to ask, what conceptualization of spatial relationships to use—all decisions that introduce subjectivity. As analysts, our focus is on making sure that we use our subject matter expertise to guide those decisions, that we are transparent about those decisions, and that we continuously assess their impact.

Also note that we have just barely scratched the surface of the discussion of p-values. It may come as a surprise to those of us outside the field of statistics, but there is increasing discussion in the scientific community about the undue weight given to p-values in research practices and decision-making.

The idea is that the p-value, in the context of statistical clustering, is just the probability of calling something a cluster when it is really just complete spatial randomness (in other words, a false positive). One of the challenges when interpreting p-values is that the probability of our null hypothesis being true impacts how we interpret the resulting p-value. For instance, we've discussed that in spatial statistics the most common null hypothesis is complete spatial randomness. Yet we also acknowledged at the beginning of the book that the first law of geography states that things that are closer together are more related than things that are farther apart. So, we don't really think complete spatial randomness is particularly likely. This gives us quite a bit of latitude when we interpret our p-values because we already expect that false positives (calling something a cluster that isn't really a cluster) aren't very likely.

If our null hypothesis was something that is very likely, like a person not having a rare genetic disease, then the chances of a false positive become higher. And our interpretation of that p-value, and what it means, varies dramatically. We probably would not feel comfortable using just a p-value from a test with a highly likely null hypothesis to make critical decisions when we know how possible a false positive is.

The good news, though, is that complete spatial randomness is not particularly likely, and the p-values that we calculate can be a useful tool for quantifying and identifying patterns in our data. So while p-values can be a bit controversial, and rightfully so, they can still be valuable if interpreted and used appropriately.

Hot spot analysis

One of the most widely used statistical clustering algorithms is the Getis-Ord Gi* statistic, which is commonly referred to as hot spot analysis. As the previous example illustrated, hot spot analysis identifies statistically significant hot spots and cold spots.

Hot spots and cold spots are areas where extreme values (very different from the mean) are concentrated.

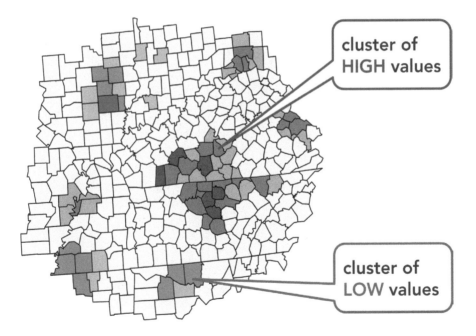

Conceptually, hot spot analysis uses the idea of neighborhoods to determine if the local average of values is significantly different from the global average (the average for the entire study area). When the average of a feature's neighborhood is very different from the global average, the feature is designated as belonging to a nonrandom cluster of values.

Let's take a look at how this works using a set of polygon **features**.

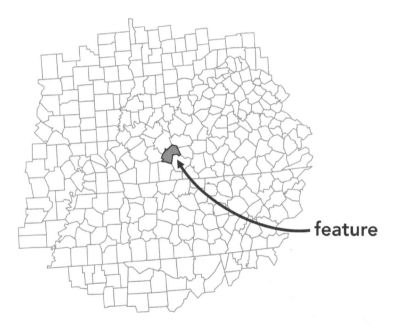

feature

We're looking for clustering in values, so each feature must have an associated attribute field with a value for us to analyze. A value could be something like a count, a rate, or any other numeric measure.

Each feature has a neighborhood that is defined by a fixed distance, a number of neighbors, or another method of conceptualizing spatial relationships.

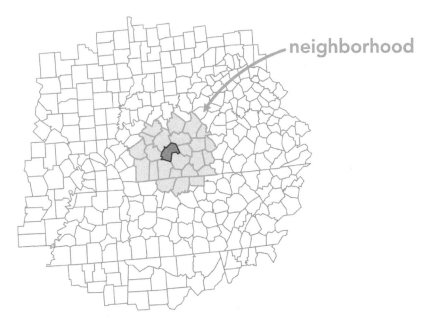

neighborhood

Hot spot analysis compares each feature's neighborhood to the entire study area to determine if that feature belongs to a cluster.

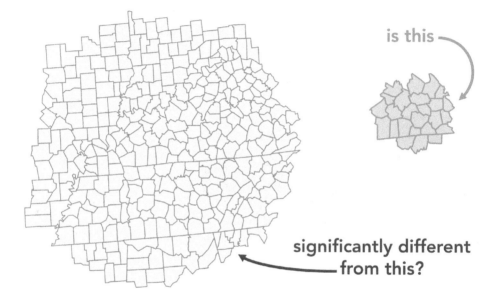

is this —

significantly different
— from this?

If the neighborhood's average is significantly higher than the global average, then the feature will be marked as a **hot spot**.

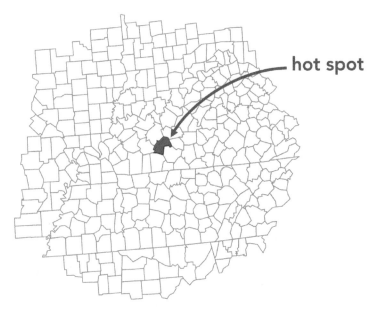

hot spot

As we discussed in the previous section, the feature's type and degree of significance are determined by its z-score and associated p-value. For a hot spot analysis, features with very positive z-scores are clusters of high values, or hot spots. Features with very negative z-scores are clusters of low values, or cold spots. A feature's p-value determines its degree of significance.

Features that belong to neighborhoods where all the values are close to the mean, or where high and low values cancel each other out, as we'd expect to see in a random distribution of values, are marked as not significant.

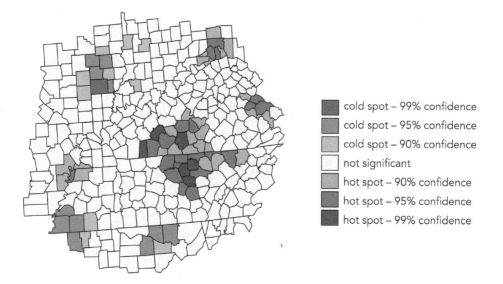

cold spot – 99% confidence
cold spot – 95% confidence
cold spot – 90% confidence
not significant
hot spot – 90% confidence
hot spot – 95% confidence
hot spot – 99% confidence

It's important to remember that the question "Where are the hot spots?" is not *necessarily* the same question as "Where are the highest values?" Since we are lumping a feature in with its neighborhood, there could be circumstances where a feature with a low value is marked as a hot spot because its neighboring values are high enough to be considered significantly different from the global average. If we just wanted to find the highest and lowest individual features, we could simply sort our table. Instead, hot spot analysis is looking for statistically significant spatial clusters.

Cluster and outlier analysis

Another really common statistical clustering technique is the Anselin Local Moran's I statistic, which we refer to as cluster and outlier analysis. It is also commonly referred to as a LISA statistic (local indicators of spatial association). The concept of cluster and outlier analysis is similar to that of hot spot analysis. Both find local clusters of high and low values, but a cluster and outlier analysis also identifies local outliers, or features that are very different from their neighbors.

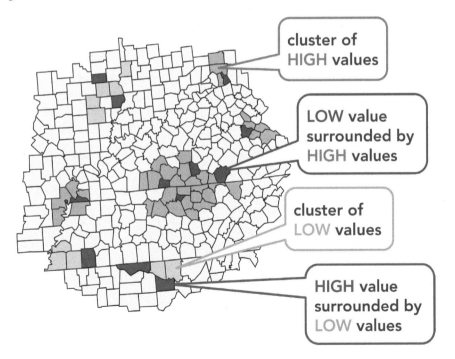

cluster of
HIGH values

LOW value
surrounded by
HIGH values

cluster of
LOW values

HIGH value
surrounded by
LOW values

As we've discussed, we expect things that are near each other to be more similar than things that are farther apart. When we find a feature that is very different from its neighbors, it is often an indicator of something worth investigating.

So how does it work? Unlike a hot spot analysis, in a cluster and outlier analysis a feature does not belong to its own neighborhood. This allows for a slightly different type of comparison. When we run a cluster and outlier analysis, we're asking two questions: 1) Is the neighborhood average significantly different from the global average, and 2) Is the feature value significantly different from the neighborhood average?

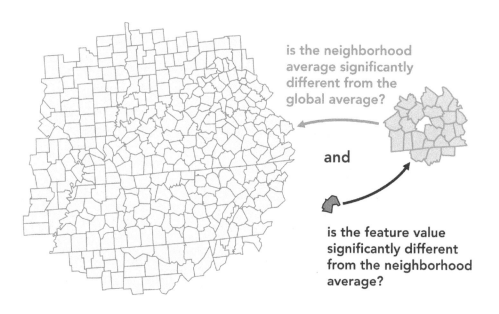

is the neighborhood average significantly different from the global average?

and

is the feature value significantly different from the neighborhood average?

Based on these questions, a feature can be categorized as one of four possible statistically significant categories.

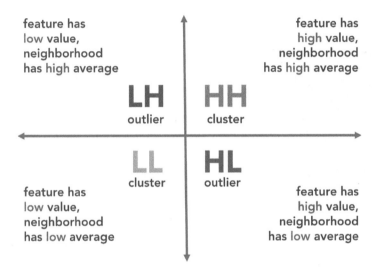

A feature with a high value surrounded by a neighborhood with a high average is categorized as a high-high cluster. A feature with a low value surrounded by a neighborhood with a low average is categorized as a low-low cluster.

Features that are very different from their neighbors are called local outliers. A feature with a high value surrounded by a neighborhood with a low average is marked as a high-low outlier. And a feature with a low value surrounded by a neighborhood with a high average is marked as a low-high outlier.

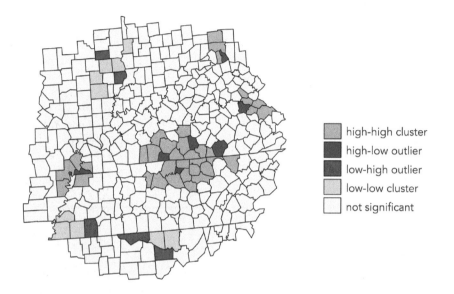

high-high cluster
high-low outlier
low-high outlier
low-low cluster
not significant

These outliers are particularly interesting and important because they are inherently spatial. A value may not be remarkable in comparison to the whole dataset, but compared to its neighbors, it may be quite remarkable. As opposed to a more traditional data "outlier," which is based on the entire dataset, they are outliers based explicitly on a comparison to their neighborhoods. This is something that would be impossible to find just looking at a spreadsheet or ignoring the spatial context of our data.

These local outliers are often locations that require further investigation. We want to understand what makes these places different from their neighbors, and this often brings up questions that we didn't even know we needed to ask.

Points and aggregation

We've learned that hot spot analysis and cluster and outlier analysis are two methods for identifying local statistical clustering of *values*. In the illustrations, we saw how these methods work with polygon data, where each polygon feature has an attribute value assigned to it. Point data can also have associated attribute values, such as the value of a home sale or the temperature at a weather station. We can analyze these in the same way we analyze polygons, where each point is evaluated in the context of its neighbors and assigned a level of significance.

There are many cases, however, when we want to analyze patterns in point data without associated attributes. For instance, these points could represent car accidents or potholes or disease occurrence. In these cases, we aren't analyzing the spatial pattern of an attribute, we're analyzing the density of the points across the study area. We create a variable that represents that density by aggregating those points.

For example, given this set of points without attribute values, we must create a density variable by aggregating them into a set of polygon boundaries so that we can count how many points occurred within each polygon.

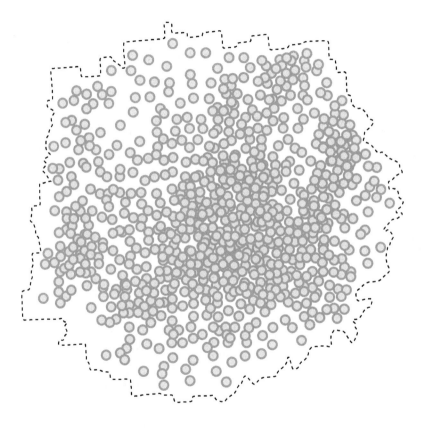

We have some options and some considerations when aggregating data.

One option is to aggregate into preexisting, standard geographies, like census boundaries.

Standard geographies can be useful when they correspond to the geographical units within which decisions will be made and actions will be taken. These kinds of polygons often also have additional associated data, such as population. This can be particularly useful in pattern analysis when we not only want to count the number of points that fall within a polygon but also create a rate using one of those additional variables.

This brings up an important question: Should we be analyzing raw counts or should we be analyzing rates? Which is more appropriate for the question that we're asking?

For instance, if we're analyzing the number of flu cases, we'd expect to have more cases in places with more people. And fewer cases in places with fewer people. So a hot spot map of the number of flu cases is very likely to look similar to a hot spot map of population. And if we're interested in where the most cases are so that we can allocate resources, then a hot spot map of those case counts is perfectly valid and useful. If, however, we're interested in finding places with more flu cases than we'd expect based on the underlying population, it would make more sense to calculate a rate. A hot spot map of flu cases per capita would reveal places where more people contracted the flu than we'd expect based on the underlying population. Using rates helps us uncover processes other than population that are causing the flu outbreak (inequitable access to health care, for instance).

One of the challenges when using standard geographies, though, is that they tend to be irregularly shaped, and their sizes can vary dramatically. This can lead to higher counts in larger polygons, which are not representative of underlying spatial processes but merely an artifact of the structure of the boundaries. Rates are one effective strategy for minimizing the impact of this variation.

Another option is to aggregate to an evenly sized fishnet or hexagon grid.

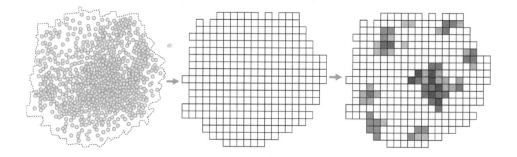

Grids can be effective because they are uniform, which leads to counts that are more representative of the underlying pattern of the initial point data. This means that we avoid some of the pitfalls of aggregating into irregularly sized polygons. We may, however, lose the wealth of additional variables that are often associated with polygonal data. Another reason we might choose to use polygons like census boundaries or our own custom boundaries is that they may be the geography at which we will remediate.

Modifiable areal unit problem

In explaining the pros and cons of using standard geographies vs. grids, we have skirted around a very important concept in geography known as the modifiable areal unit problem, or MAUP. The "problem" is that aggregating data has an impact on interpretation of patterns, and different aggregation schemes can, and often do, lead to different results and different answers. In other words, the shape and size of the polygons that we aggregate into can greatly impact our results.

One important aspect of MAUP is known as the scale effect.

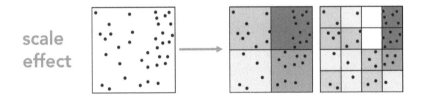

Simply changing the size of the polygons that we are aggregating into can dramatically change the results. For instance, if we choose a very large grid cell size, we may hide important clusters in our data that are entirely subsumed within large grids.

Another aspect of MAUP is known as the zone effect.

In the case of the zone effect, changing the shape of the polygons that we are aggregating into can dramatically change the results. The zone effect is important to consider when aggregating into artificial boundaries like census boundaries or other administrative units.

How does this relate to hot spot analysis? It's important to recognize that any time we aggregate our data, we are imposing a structure on it. We want our modeling to pick up the data itself, not the artificial structure we've imposed. Using an aggregation scheme that reflects the pattern (and spatial processes creating that spatial pattern) will diminish the impact of MAUP.

Learning about MAUP can be a little overwhelming because it can seem as if there is no good way to aggregate our data, and the truth is that each choice does have pros and cons that need to be understood and evaluated. When choosing an appropriate grid size, for instance, we must think about the scale of the question we're asking and how that relates to the extent and density of the points being aggregated. Very dense points will likely require smaller grid cells than those that are sparely distributed in the same spatial area. We can evaluate the impact that our chosen grid size has on the resulting spatial patterns by trying a number of different cell sizes and seeing how sensitive the hot and cold spots are to the changing cell sizes.

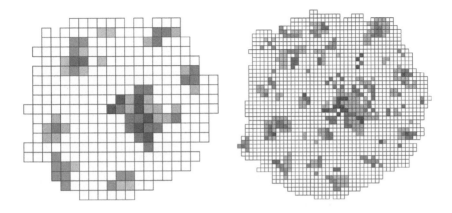

Zeros vs. nulls

Something else to consider when aggregating is the very important difference between zero and null. When aggregating, we need to decide what it means when a polygon or a cell is empty. In some cases, an empty cell means a count of zero. In other cases, a cell might be empty because it is not possible for an event to have occurred there, meaning a null is appropriate.

For example, if we're counting the number of house fires within each grid cell, it seems safe to assume that if no points fall within a cell, it represents zero fires. But what if that grid cell falls in the middle of a river? Is that zero meaningful? Probably not. It is important to only aggregate into cells where house fires could truly have occurred, to ensure that all zeros in the dataset are meaningful.

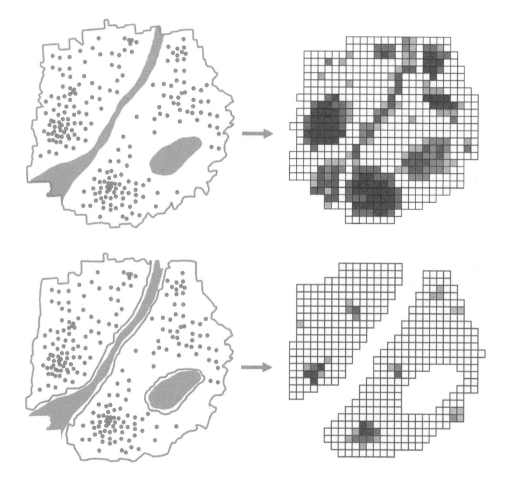

This is particularly important because both hot spot analysis and cluster and outlier analysis use averages, and having a lot of zeros in a dataset can impact those averages dramatically. If the dataset is inflated with a lot of zeros, it can pull down the global average and make everything seem like a cluster. For this reason, limiting our analysis features to only a meaningful study area is critical. For instance, if we know that house fires are not possible in water bodies, a grid used for aggregation should omit all cells that fall within water bodies.

What's the "right" statistical clustering method?

So, which one of these statistical clustering methods should we use? The truth is, each method takes a different approach to finding clusters in our data. They have strengths and weaknesses. And as unsatisfying as this answer may seem, our best advice is to give them both a try. Each algorithm we apply to our data, each time we analyze it through a slightly different lens, we learn something new about the underlying spatial processes at play. When the results line up, we feel even more confident in the patterns that we're seeing. When there are discrepancies, we dig deeper and learn more.

Chapter 5

Spatiotemporal pattern mining

Introduction

So far, we've talked about finding patterns in space. But what about when our data has time? In the same way that ignoring the spatial dimension of our data would give us an incomplete picture, leaving out the temporal dimension is a missed opportunity to gain a deeper, more nuanced understanding.

There are a number of ways that we can incorporate time into our analysis. One of the most common ways is to analyze snapshots of data. For instance, we could take a year of unemployment data and look at it month by month, either as thematic maps or even as hot spot maps for each month, to see how things are changing over time.

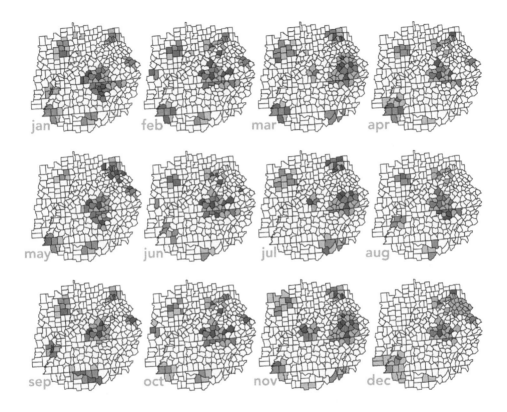

One of the challenges with this approach, where we have to look at multiple maps one after the other, is that it can be difficult to visually quantify things like trends and change. This approach also assumes that each time period exists completely independently. To analyze February without taking into account what happened in January leaves important information out of our analysis.

We learned in the last chapter that statistical cluster analysis uses the concept of neighborhoods to find patterns in our data. So far, we've defined those neighborhoods in terms of what's near in space. In this chapter, we'll explore what's possible when we expand our definition of near to include not just space but also time.

We're going to show you how Tobler's first law of geography can be extended to include temporal relationships: *Everything is related to everything else, but near and recent things are more related than distant things.* And we'll explore how we can apply the statistical clustering methods we learned about in the last chapter to spatiotemporal data.

Methods in this chapter can help us answer the following types of questions:

- What are the spatiotemporal trends in flu cases?

- How have concentrations of air pollution changed since stricter regulations took effect?

- How have clusters of unemployment changed over time, and which areas are getting worse?

Space-time cubes

To analyze spatiotemporal data, we need a common data format to simplify and standardize our analysis. With hot spot analysis and cluster and outlier analysis, that data format is features with associated values. For spatiotemporal data, that format has to be features with values that are measured over time. Sometimes our data is collected in this way—for example, counties collecting unemployment rates over time. Sometimes our data has to be aggregated because it is point data with locations and times, but no associated values. Whether we already have a more traditional time series dataset or a point dataset that requires spatiotemporal aggregation, we can use a space-time cube as a common data format.

The anatomy of a space-time cube

Conceptually, it can be helpful to imagine the cube in 3D, where features at the top have the most recent values. The cube can be structured as a grid or as a series of repeating irregular polygons.

grid aggregation

polygon aggregation

Each block of space and time within the space-time cube is called a **bin**.

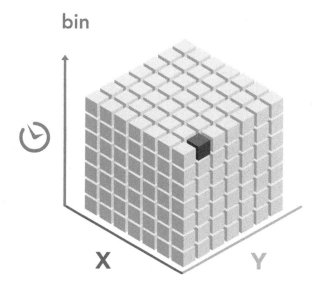

Bins are the individual units that make up the space-time cube, each with a unique spatio-temporal extent.

A column of bins that share the same spatial extent is called a location.

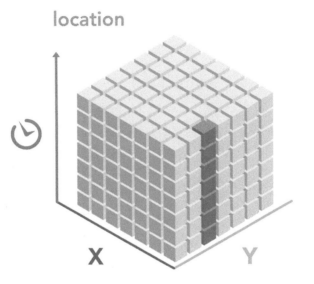

These bins share a location but have different temporal extents. Each location is essentially a time series; when we create the cube, we end up with time-series information for each location.

The rows that share a temporal extent are called a time slice.

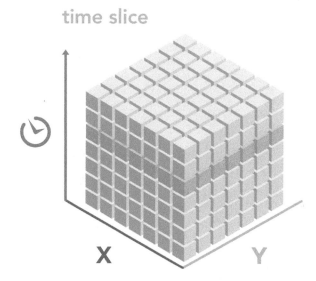

These time slices, in many respects, are the way that we've always done temporal analysis in GIS. They are snapshots, so we can compare one snapshot to the next snapshot over time, which is valid and useful. However, in doing it that way, we are making the claim that each time slice is independent and exists in a vacuum; it has nothing to do with what came before it, and it has nothing to do with what comes after it. Space-time pattern mining tools integrate time in a tighter way, where the time slice is being interrelated in a 3D structure.

Aggregating points into space-time cubes

Point data must first be aggregated into a space-time cube in order to analyze it using these spatiotemporal clustering techniques. This process is very similar to the aggregation we've already discussed, with a few additional considerations. When aggregating a set of points spatially, we count the number of points that fall within a polygon. When aggregating into a space-time cube, we're counting the number of points that fall within each bin, which are defined not only by their spatial extent but also their temporal extent.

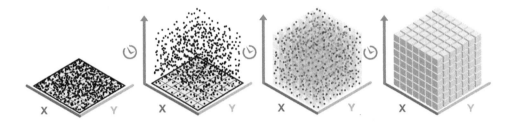

Each point falls into a bin based not just on where it happened but also *when* it happened.

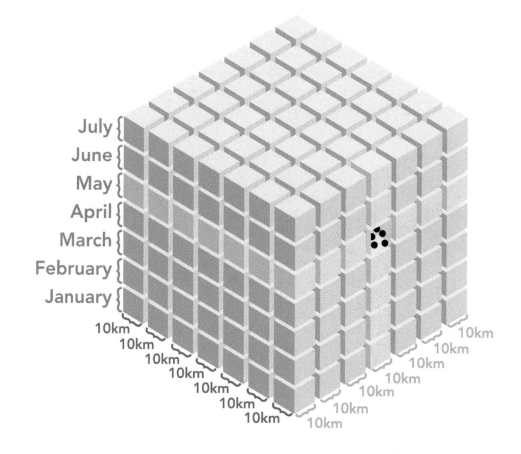

Modifiable temporal unit problem

Aggregating time can be tricky. The good news is that our experience aggregating things in space actually prepares us well to think through aggregating in time.

We've already discussed the impact that spatial aggregation can have on our analysis in the form of the modifiable areal unit problem. In addition to thinking about the impact of our spatial aggregation, we also have to think about the impact of our temporal aggregation. How we divide the data temporally (weekly, 30 days, etc.) can dramatically impact the results of our analysis. This is sometimes referred to as the modifiable temporal unit problem. For example, an analysis that looks at monthly data will yield very different results from the same data aggregated to years. All seasonality, for instance, will be hidden in the yearly aggregation, but more coarse patterns may become more apparent. Another example is the use of months in general, which vary in length from 28 to 31 days. If we're counting the number of car accidents per month, February may always seem like the safest month just because it is the shortest month of the year.

One less obvious example of this problem is something that we refer to as temporal bias.

Let's use this dataset to illustrate.

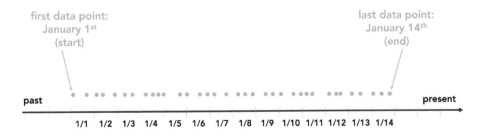

Imagine we wanted to aggregate this data into four-day chunks. We could start with the most recent data point and count backwards.

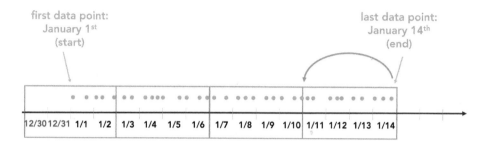

But you'll notice that the bin in the beginning of the dataset is half empty.

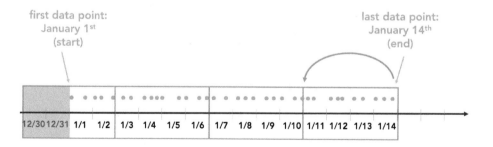

We'd encounter the same issue at the other end of the dataset if we began aggregation from the earliest data point.

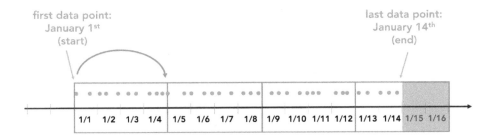

Making sure that our temporal aggregation divides evenly among our data will help reduce temporal bias and ensure results that are representative of the patterns in our data and not artifacts of our aggregation.

Emerging hot spot analysis

Once our data is in a space-time cube, we're ready for analysis.

Emerging hot spot analysis is the spatiotemporal extension of hot spot analysis. It uses the same Getis-Ord Gi* statistic but extends the concept of what it means to be a neighbor to include not only what is near in space but also what is recent in time.

Within the space-time cube, each bin is evaluated in the context of its neighboring bins. This includes the bin's spatial neighbors, the ones that are closest geographically, and its temporal neighbors, the ones that are closest in time, as well. Only bins that are both proximate in space and recent in time are considered related.

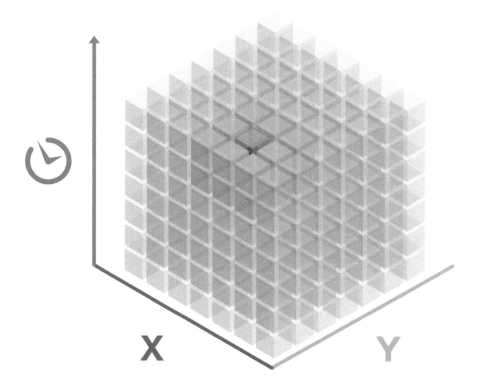

With this three-dimensional conceptualization of proximity, each bin's neighborhood is defined and then compared to the study area. If the neighborhood value is significantly higher than the study area, then that bin is marked as a hot spot.

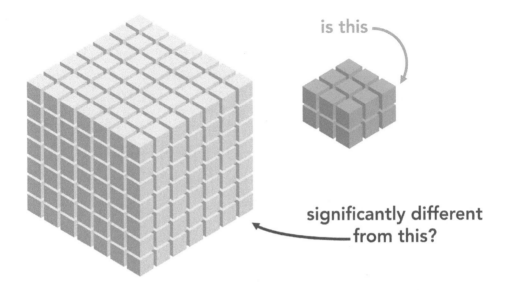

is this

significantly different
from this?

Just like in a two-dimensional hot spot analysis, each bin is assigned a probability that quantifies how likely it is to belong to a nonrandom cluster of high values, a hot spot, or a non-random cluster of low values, a cold spot.

To interpret the results of an emerging hot spot analysis, we use both a 3D result and a 2D result. The 3D result is useful as it gives us the statistical significance of clustering at each bin over time for the entire space-time cube, but it can be a lot of information to interpret on its own. To facilitate interpretation, the type and intensity of clustering is then summarized for each location and categorized based on the location's pattern or trend in clustering over time.

There are 17 possible categories, eight for hot, eight for cold, and one for no pattern. Each category describes a unique temporal pattern.

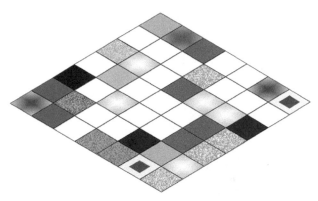

new hot spot
consecutive hot spot
intensifying hot spot
persistent hot spot
diminishing hot spot
sporadic hot spot
oscillating hot spot
historical hot spot
new cold spot
consecutive cold spot
intensifying cold spot
persistent cold spot
diminishing cold spot
sporadic cold spot
oscillating cold spot
historical cold spot
no pattern detected

For example, this location has been marked as a Sporadic Hot Spot, meaning that most recently it was hot, but over time it has switched back and forth between hot and not significant.

 sporadic hot spot

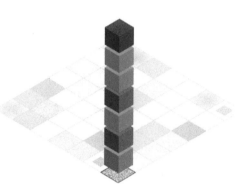

This location is marked as an Intensifying Hot Spot, meaning that it has been hot at least 90% of the time and that a significant upward trend is detected in the clustering intensity. Essentially, the hot spot is getting hotter.

 **intensifying hot spot**

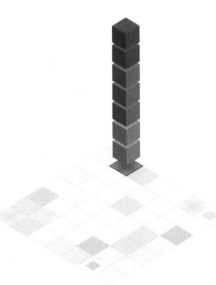

And this location is marked as a New Hot Spot, meaning that it had never been hot before, until the most recent time period, when it became hot for the first time.

 new hot spot

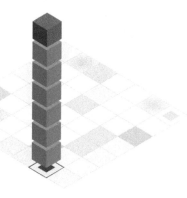

Each of the categories is described in the table.

Category	Definition
No Pattern Detected	Does not fall into any of the hot or cold spot patterns defined below.
New Hot Spot	A location that is a statistically significant hot spot for the final time step and has never been a statistically significant hot spot before.
Consecutive Hot Spot	A location with a single uninterrupted run of statistically significant hot spot bins in the final time-step intervals. The location has never been a statistically significant hot spot prior to the final hot spot run, and less than 90% of all bins are statistically significant hot spots.
Intensifying Hot Spot	A location that has been a statistically significant hot spot for 90% of the time-step intervals, including the final time step. In addition, the intensity of clustering of high counts in each time step is increasing overall, and that increase is statistically significant.
Persistent Hot Spot	A location that has been a statistically significant hot spot for 90% of the time-step intervals with no discernible trend indicating an increase or decrease in the intensity of clustering over time.

Diminishing Hot Spot	A location that has been a statistically significant hot spot for 90% of the time-step intervals, including the final time step. In addition, the intensity of clustering in each time step is decreasing overall, and that decrease is statistically significant.
Sporadic Hot Spot	A location that is an on-again then off-again hot spot. Less than 90% of the time-step intervals have been statistically significant hot spots, and none of the time-step intervals have been statistically significant cold spots.
Oscillating Hot Spot	A statistically significant hot spot for the final time-step interval that has a history of also being a statistically significant cold spot during a prior time step. Less than 90% of the time-step intervals have been statistically significant hot spots.
Historical Hot Spot	The most recent time period is not hot, but at least 90% of the time-step intervals have been statistically significant hot spots.
New Cold Spot	A location that is a statistically significant cold spot for the final time step and has never been a statistically significant cold spot before.
Consecutive Cold Spot	A location with a single uninterrupted run of statistically significant cold spot bins in the final time-step intervals. The location has never been a statistically significant cold spot prior to the final cold spot run, and less than 90% of all bins are statistically significant cold spots.

Intensifying Cold Spot	A location that has been a statistically significant cold spot for 90% of the time-step intervals, including the final time step. In addition, the intensity of clustering of low counts in each time step is increasing overall, and that increase is statistically significant.
Persistent Cold Spot	A location that has been a statistically significant cold spot for 90% of the time-step intervals with no discernible trend indicating an increase or decrease in the intensity of clustering of counts over time.
Diminishing Cold Spot	A location that has been a statistically significant cold spot for 90% of the time-step intervals, including the final time step. In addition, the intensity of clustering of low counts in each time step is decreasing overall, and that decrease is statistically significant.
Sporadic Cold Spot	A location that is an on-again then off-again cold spot. Less than 90% of the time-step intervals have been statistically significant cold spots, and none of the time-step intervals have been statistically significant hot spots.
Oscillating Cold Spot	A statistically significant cold spot for the final time-step interval that has a history of also being a statistically significant hot spot during a prior time step. Less than 90% of the time-step intervals have been statistically significant cold spots.
Historical Cold Spot	The most recent time period is not cold, but at least 90% of the time-step intervals have been statistically significant cold spots.

These categories can be useful tools for interpretation, but they're also not the be-all and end-all of how to understand the results. They are just one of many ways that these trends could be categorized. The p-values and z-scores from the Getis-Ord Gi* statistic are concrete, but the categories from emerging hot spot analysis are an interpretive tool. Depending on the audience or the question that is being asked, it may be appropriate to combine categories or even come up with new ones. For instance, our audience may not care about the difference between an intensifying, diminishing, or persistent hot spot. It may be enough that they have all been significant for over 90% of the time, and they can be grouped accordingly.

Local outlier analysis

Local outlier analysis extends the Anselin Local Moran's I statistic to identify value clusters and detect local outliers. Local outlier analysis also uses a spatiotemporal conceptualization of what it means to be a neighbor to identify statistically significant clusters and outliers in the context of both space and time.

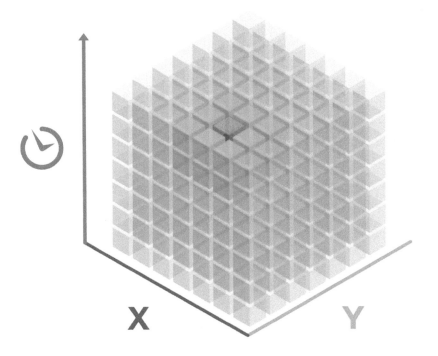

With this three-dimensional conceptualization of proximity, each bin is compared to its neighborhood average, and each neighborhood average is compared to the study area (exactly the same questions that we ask with the two-dimensional cluster and outlier analysis).

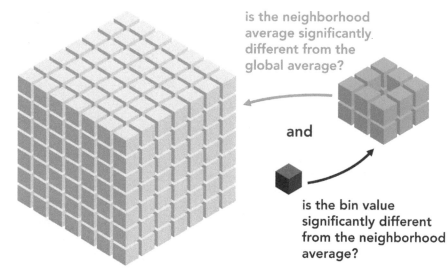

is the neighborhood average significantly different from the global average?

and

is the bin value significantly different from the neighborhood average?

Just like the two-dimensional cluster and outlier analysis, the result includes four possible statistically significant categories.

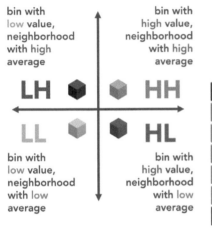

bin with low value, neighborhood with high average

bin with high value, neighborhood with high average

LH

HH

LL

HL

bin with low value, neighborhood with low average

bin with high value, neighborhood with low average

Bins can be categorized as either high or low clusters or as high or low outliers within clusters. Every bin in the 3D cube has an associated category.

Just like with emerging hot spot analysis, a summary output of local outlier analysis is critical for interpretation. This summary tells us if a location has ever been significant, and if so, which type it is.

only high-high cluster

only high-low outlier

only low-high outlier

only low-low cluster

multiple types

never significant

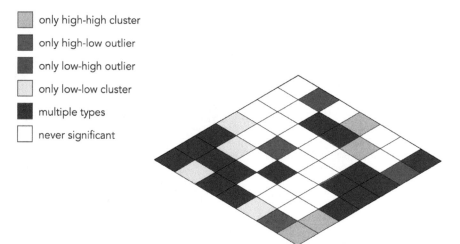

For example, this location has been marked as Only High-High Cluster, because over time it has been significant, and only as a High-High Cluster.

 only high-high cluster

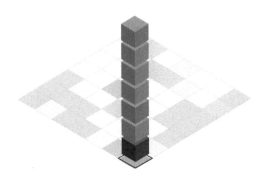

This location has been marked as Multiple Types, because over time it has been significant as both a High-Low Outlier and a Low-Low Cluster.

 multiple types

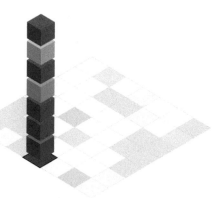

The categories for local outlier analysis do not quantify trends or intensity of clusters or outliers over time. For instance, a location with a single high outlier and a location that is a high outlier for the entire time period will be categorized the same way. As a result, investigating areas of interest using the 3D cluster and outlier results is important for understanding these patterns and trends. Finding a location that has flipped between being a high outlier and a cold spot in a cyclical way over time, for example, can help us uncover underlying factors that contribute to these repeating patterns.

Summary

It can be challenging to incorporate time into our spatial analysis, but it can also provide tremendous value when done right. In this chapter, we've learned how to structure our spatiotemporal data effectively and the importance of accounting for temporal bias. We've explored the ways that we can extend common statistical clustering methods to incorporate time. And we've illustrated how to overcome the challenges of visualizing and interpreting these complex multidimensional results with techniques like categorization and 3D visualization.

Moving beyond a purely spatial analysis allows us to uncover not only spatial patterns but also temporal trends in our data. These trends are powerful because they help us quantify change, evaluate the impacts of our actions, and anticipate what's coming.

Chapter 6

Modeling spatial relationships and making predictions

Introduction

So far, we've learned several different ways that we can investigate our spatial data to identify areas with meaningful patterns. After we've identified where the pattern is, the next logical question is "What's behind that pattern?" We can use modeling techniques to find relationships that help us explain, predict, and prescribe.

So what's a model? The idea behind a model is that it represents reality in some way that will be useful to us. It's an abstraction that we can use to make predictions. We use models for all sorts of things in the world. Role models, model airplanes, fashion models…all of these act as archetypes that represent some important facet of what is being "modeled," and they can be valuable when building up our understanding of how things work. The same can be said for statistical models. They capture relationships that exist in our data so we can use information that we have to predict information that we don't have. For example, if we can find the relationship between altitude and temperature, we can predict temperature based on altitude, or vice versa.

Now, what does it mean for variables to be related to each other? There are many different types of relationships between variables, but in the simplest sense, variables are related if information can be learned about one variable by observing values of other variables. So since altitude and temperature are related, information can be gained about temperature by observing information about altitude. This is called dependence between two variables. Conversely, if no information can be gained about one variable by observing the other variable, then the variables are independent. Statistical modeling is all about modeling these often complex relationships so that we can understand what's behind the patterns that we're seeing and make predictions based on that understanding.

All of that said, we do have to be careful. Some models are more useful than others. There are a lot of ways that we can end up with a model that is lousy at predicting. We want a model that is not too simple but also not too complex; it's a classic Goldilocks situation.

When a model is too complex, we run the risk of overfitting. A model is overfit when it's so specifically tailored to the data that we use to train the model that it cannot generalize, and if our model cannot generalize, it cannot predict.

When a model is too simple, we run the risk of underfitting. A model is underfit when it's so general that it doesn't capture the relationships in the training data and is therefore incapable of generating useful predictions.

The "just right" model is one that captures the relationships, but not so closely that it cannot generalize. There are a number of important ways that we can evaluate the fit and performance of our model, and we'll talk more about that throughout the chapter.

underfit   good fit   overfit

There are countless ways to approach modeling, from linear methods all the way to deep learning. We could write a whole book just about modeling (many have!). Here, we're going to focus on two linear regression methods: ordinary least squares (OLS) regression and geographically weighted regression (GWR). OLS is one of the most widely used statistical modeling techniques around. Even though the world of modeling is evolving rapidly in terms of both statistical and machine learning–based approaches, understanding OLS provides a critical foundation as we enter this space. We will then build on our understanding of OLS, which is a nonspatial method, and explore its spatial counterpart, GWR.

Methods in this chapter can help us answer the following types of questions:

- What areas are most prone to mosquito-borne illness?

- How do the factors that influence home values change across space?

- What factors contribute to high Medicare spending? What remediations would have the most impact and where?

Ordinary least squares regression

Ordinary least squares is a linear regression method, which means that the outcome of the model is an equation that describes the form of a line.

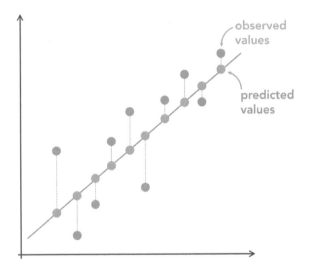

observed values

predicted values

You can think of that line as the prediction, where we use the line to determine a value for any new data that we want to predict. To understand this, let's look at the equation (model) that defines the line:

$$y = \beta_0 + \beta_1 X_1 + \beta_2 X_2 + \dots \beta_n X_n + \varepsilon$$

Don't be intimidated. We're just going to use this as a framework to go over the terminology that is necessary to understand what's happening under the hood in OLS.

Starting on the left side of the equation, **y** is our dependent variable.

$$y = \beta_0 + \beta_1 X_1 + \beta_2 X_2 + \ldots \beta_n X_n + \varepsilon$$

Our dependent variable is what we're interested in modeling, predicting, and understanding. For example, if we want to predict precipitation, then precipitation is our dependent variable.

The Xs are the explanatory variables.

$$y = \beta_0 + \beta_1 X_1 + \beta_2 X_2 + \ldots \beta_n X_n + \varepsilon$$

These are the variables that we believe explain, or have a relationship with, the dependent variable.

These are also called independent variables, but we prefer the term *explanatory variables* because it's more descriptive.

In our example of modeling precipitation, we might use explanatory variables like altitude, temperature, and humidity. So, looking at the equation, precipitation equals some combination of these explanatory variables multiplied by a number, also known as a coefficient.

Each explanatory variable has a **coefficient**.

$$y = \beta_0 + \beta_1 X_1 + \beta_2 X_2 + \ldots \beta_n X_n + \varepsilon$$

Coefficients represent the strength and type of relationship that the explanatory variable has to the dependent variable when holding all other explanatory variables constant.

Coefficients can have a positive relationship (as X rises, so does y), a negative relationship (as X rises, y drops), or no relationship (X is not correlated to y).

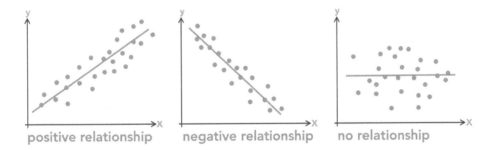

positive relationship   negative relationship   no relationship

Coming back to our example, we know how temperature, altitude, and humidity relate to precipitation based on their coefficients. They are each multiplied by their coefficient in the equation to create the prediction.

Last, we have our **residuals.**

$$y = \beta_0 + \beta_1 X_1 + \beta_2 X_2 + \ldots \beta_n X_n + \varepsilon$$

Residuals are the error term. They are the model's over- and underpredictions.

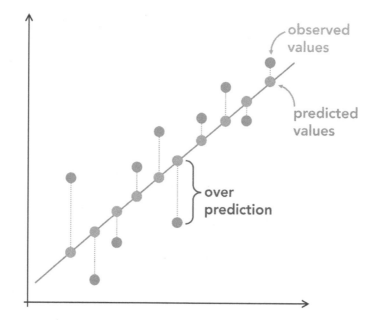

In the simplest example with a single explanatory variable, you can think of a linear model as a prediction line. This visual is useful for understanding the concept, but the true representation of a multivariate model (a model with multiple explanatory variables) would be multidimensional and significantly more difficult to illustrate. So, back to the line. We compare where our observed values fall in relation to that prediction line. The difference between the observed value and the predicted value is a residual. These residuals are important because this difference tells us what wasn't represented or what wasn't explained with the model that we've created.

And it's really that simple. This equation defines the linear regression model, and that model is used to explore and understand relationships as well as make predictions.

Coming back to our example, once we have a prediction model for precipitation, if we get new data where we only know temperature, altitude, and humidity, we can use the equation to calculate precipitation.

Geographically weighted regression

Now that we have OLS as a foundation, we can make it spatial! And that's just what geographically weighted regression does. The big difference between OLS and GWR is that OLS is a global model. That means there is one single equation for the entire study area based on every single feature in the study area. All the data is used to calibrate the model. GWR, however, is local. It's spatial. With GWR, each feature in the study area gets its own separate regression equation that is calibrated based on the feature's neighborhood.

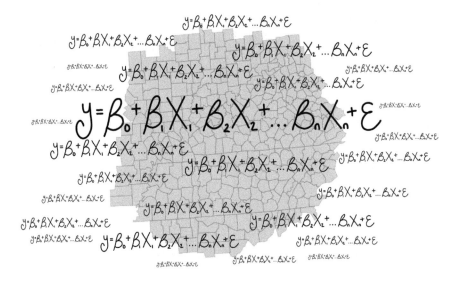

One of the biggest benefits of this is that we can improve our model by acknowledging that nearby features are more related than distant features (that first law of geography just won't quit!). We can use the same types of neighborhood definitions that we've discussed throughout the book.

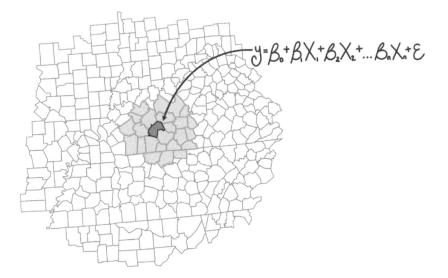

By calibrating a model for each feature based on its neighbors, we're able to capture the spatial variation in the relationships in our data. As a result, since we have individual models for each feature, the coefficients for each explanatory variable are allowed to vary across the study area.

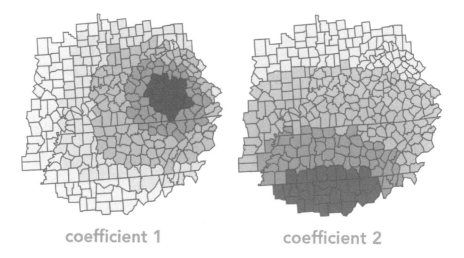

coefficient 1   coefficient 2

Let's think about this using an example—for instance, modeling home values using distance from amenities (like supermarkets or Laundromats). In dense, urban areas, where most travel happens on foot or using public transportation, being close to these amenities may raise home values. On the other hand, in suburban areas where much of the travel happens by car, we might find that being too close to these types of amenities is undesirable, lowering home values. The exact same variable can not only vary in importance, but sometimes the direction of the relationship can even flip (being close to a supermarket is positive in some places and negative in other places).

And allowing for that variation helps us make better predictions and understand these complex phenomena in a much more nuanced way. Knowing where an explanatory variable has the most impact on the dependent variable can help us tailor our remediation to focus on those areas where it will be most effective.

Model evaluation

As we discussed earlier, not all models are good models. There are quite a few things to consider and diagnostics to evaluate to determine if our model is a good one.

Explanatory variables

For one thing, we want to be thoughtful when choosing explanatory variables. They should each have a reason for being included. The more explanatory variables we include, the more complex our model becomes. Our goal is for the model to be as simple as possible, but no simpler (a concept often referred to as parsimony). To evaluate the importance of each variable in the model, statistical significance is calculated for each corresponding coefficient. Including only variables with significant coefficients is an effective way to ensure we're using only variables that are adding value to our model.

When our goal is not just to predict but to explain or understand the phenomenon we're modeling, we also want to make sure that each variable we include is unique in the model. When two variables included in the model are telling the same story, it leads to something called multicollinearity. This just means that the variables are redundant, or highly correlated to each other. If we were modeling house prices, for instance, we would likely want to include a variable that gets at the size of the house. It might be tempting to include both square footage and number of bedrooms, but in many instances these variables are redundant and can cause issues in the model. A common diagnostic that evaluates multicollinearity is the variance inflation factor, or VIF. High VIFs indicate that we have issues with multicollinearity in our model and should consider including just one of the redundant variables.

Residuals

Model residuals are one of the most important indicators that we have to evaluate our model. We want our residuals to have a random spatial pattern, and we want them to be normally distributed with a mean of zero. Let's explore why.

Spatial clustering of our residuals indicates that our model might be good at predicting in one part of our study area, but not in another part of our study area. Ideally, we want our residuals to look like a random spatial pattern with no clustering of our over- and underpredictions. If all the overpredictions were happening on one side of our study area and all underpredictions were happening on the other side of our study area, it might indicate that we are missing key explanatory variables in our model that capture the differences in the characteristics across the study area. For instance, if we were modeling precipitation and failed to include elevation, we might expect the model to underpredict in mountainous areas and overpredict in flat areas.

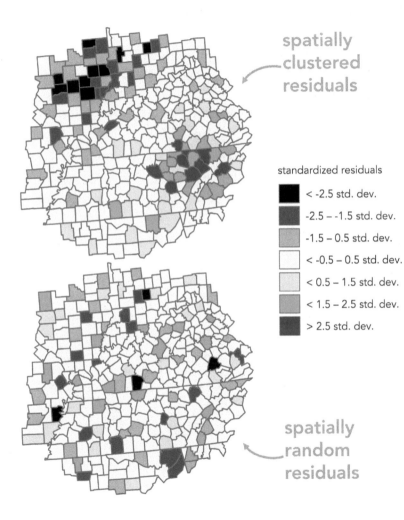

spatially clustered residuals

standardized residuals

■ < -2.5 std. dev.

■ -2.5 – -1.5 std. dev.

■ -1.5 – 0.5 std. dev.

□ < -0.5 – 0.5 std. dev.

■ < 0.5 – 1.5 std. dev.

■ < 1.5 – 2.5 std. dev.

■ > 2.5 std. dev.

spatially random residuals

We can measure clustering of our residuals using a technique that measures spatial auto-correlation called Moran's I. Moran's I is a global spatial statistic that measures the correlation between the values of features and the distances between those features. There is cluster-ing when similar values are close together and different values are far apart. So, for residu-als, we'd have clustering if overpredictions are all close together and underpredictions are all close together. Spatial autocorrelation is a key diagnostic to evaluate residuals when model-ing spatial data.

We also want our residuals to be normally distributed with a mean of zero.

In an ideal model, we would expect that over- and underpredictions are close to zero and taper off equally on either side, creating a normal distribution in the shape of a bell curve.

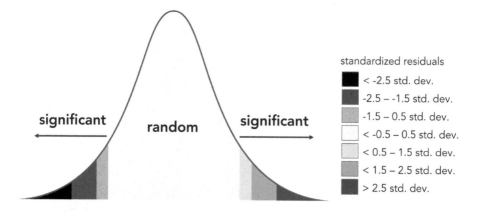

In addition to visualizing our residuals in a histogram, one measure we can use to evaluate if our residuals are normally distributed is the Jarque-Bera statistic.

If the model returns a statistically significant Jarque-Bera statistic, then our residuals are not normally distributed and our model requires further tuning.

R-squared

We have one more model evaluation diagnostic to discuss, and it is the most well-known and widely used diagnostic for linear regression: R-squared (R^2). R^2 is a measure that quantifies how much variance in our dependent variable can be explained by our explanatory variables. Another way of thinking about it, in the context of the simplest model with a single explanatory variable, is measuring how close the actual values are to the prediction line. The higher the R^2, the better the model fits the line.

Identifying a "good" model is subjective and varies greatly based on the field of study and how the model will be used. In many fields, including many of the social sciences, an R^2 value over 0.70 might be considered satisfactory for making a prediction. In drug trials, however, R^2 requirements would be much higher.

Correlation vs. causation

Models have a variety of uses. Sometimes we want to create a model for the sole purpose of making accurate predictions in places where we don't have data. Sometimes we create a model to explain a complex phenomenon so that we can remediate or take action to change outcomes. Especially when our goal is to understand underlying factors of a complex phenomenon, we have to be careful about how we interpret and use our model.

You've probably heard the phrase "correlation does not imply causation," and nowhere is that more true than when doing statistical modeling. Just because a series of explanatory variables can effectively predict a dependent variable does not mean that they are the cause of that dependent variable. Or that changing one of them would impact the dependent variable. For instance, we might find that ice cream sales are a good predictor of pool pass sales. This is useful if all we want to do is predict pool pass sales based on ice cream sales. But if we want to increase pool pass sales, we will likely not accomplish that by changing ice cream sales. The reality is, there is a confounding variable, temperature, at play. These two variables are absolutely correlated, but assuming causation would be problematic.

It seems fitting to end our modeling chapter with one of our favorite quotes.

"All models are wrong, but some are useful." —George E. P. Box

It's a nice reminder that our models are never going to be perfect, but if we're careful when we build and interpret them, they can help us understand, predict, and prescribe effectively.

Conclusion

It goes without saying that we are only scratching the surface of what's possible. There is a whole world of spatial statistics out there. We've chosen some of the fundamental techniques to explore in the hope that they will both help you in your analyses and provide a foundation as you explore that broader world of spatial statistics.

Understanding these methods isn't just about being able to use them; it's also critical for the next step in your analysis, which is communicating the results and why they matter. We can do the best analysis that has ever been done in the history of the world, but if we can't effectively communicate it, it's not useful. The good news is, with an understanding of how these methods work, you're well prepared to explain what you did, what you found, and what it means.

And, of course, we don't solve big, complex problems using just a single method or even just spatial statistics. We don't want to be a hammer walking around looking for a nail. As we discussed in the beginning, breaking down complex problems into answerable questions is a key first step in our analysis. And these methods that we've explored will be valuable tools in your toolbox, but they certainly won't be the only ones.

Our greatest hope in writing this book is that you will walk away feeling empowered and inspired to use these methods to do great analysis.

Welcome to the beautiful world of spatial statistics.

Acknowledgments

First and foremost, we'd like to thank the entire spatial statistics team, past and present, for their hard work creating and implementing these powerful and approachable tools in ArcGIS. We'd especially like to thank Eric Krause, Kevin Butler, Orhun Aydin, Alberto Nieto, and Mark Janikas for their thorough reviews and guidance as we brought this book to life.

None of this would have been possible without Lauren Scott Griffin, who not only built the first spatial statistics tools in ArcGIS but has also been a constant mentor and inspiration to both of us throughout our careers. This book is as much yours as it is ours.

Thank you to the amazing folks at Esri Press, especially Claudia Naber for helping us get over the finish line with her endless patience and support. Thank you to Clint Brown for giving us the latitude to work on a project like this. And thank you to Jack and Laura Dangermond for creating a place where we could find our passion and run with it.

Last, we'd like to give some personal acknowledgments.

From Flora:

Thank you to Cory and Billy for being my cheerleaders and seeing me through this process. Thank you to my mom for filling my life with opportunities. And mostly, thank you, Lauren. Writing this book with you has been a highlight of my life, and there is literally no one else on the planet that I could have done this with (or would have wanted to).

From Lauren:

Thank you to Nate for always supporting me when I take on at least one more project than I should. Thank you to George and Theresa for somehow simultaneously making it feel like there aren't enough hours in the day and also like anything is possible. And thank you, Flora. For everything.

References

Chapter 1: Why spatial is special

Tobler W. R. (1970). A computer movie simulating urban growth in the Detroit region. *Economic Geography*, 46, 234–240. https://doi.org/10.2307/143141.

Getis, A., & Aldstadt, J. (2004). Constructing the spatial weights matrix using a local statistic. *Geographical Analysis*, 36(2), 90–104. https://doi.org/10.1353/geo.2004.0002.

Mitchell, A. (2005). *The Esri guide to GIS analysis, vol. 2.* Esri Press.

Chapter 2: Means and medians

Mitchell, A. (2005). *The Esri guide to GIS analysis, vol. 2.* Esri Press.

Burt, J. E., & Barber, G. (1996). *Elementary statistics for geographers.* Guilford.

Kuhn, H. W., & Kuenne, R. E. (1962). An efficient algorithm for the numerical solution of the Generalized Weber Problem in spatial economics. *Journal of Regional Science*, 4(2), 21–33.

Wang, B., Shi, W., & Miao, Z. (2015). Confidence analysis of standard deviational ellipse and its extension into higher dimensional Euclidean space. *PloS ONE*, 10(3), e0118537.

Chew, V. (1966). Confidence, prediction, and tolerance regions for the multivariate normal distribution. *Journal of the American Statistical Association*, 61(315), 605–617.

Mardia, K.V., & Jupp, P. E. (2000). *Directional statistics.* John Wiley & Sons.

Chapter 3: Finding clusters with machine learning

Birant, D. & Kut, A. (2007). ST-DBSCAN: An algorithm for clustering spatial–temporal data. *Data & Knowledge Engineering*, 60(1), 208–221. https://doi.org/10.1016/j.datak.2006.01.013.

Ester, M., Kriegel, H. P., Sander, J., & Xu, X. (1996). A density-based algorithm for discovering clusters in large spatial databases with noise. *kdd*, 96(34), 226–231.

Campello, R. J., Moulavi, D., & Sander, J. (2013). Density-based clustering based on hierarchical density estimates. In *Advances in Knowledge Discovery and Data Mining: 17th Pacific-Asia Conference, PAKDD 2013, Gold Coast, Australia, April 14–17, 2013, Proceedings, Part II 17* (pp. 160–172). Springer Berlin Heidelberg. https://hdbscan.readthedocs.io/en/latest/how_hdbscan_works.html.

Agrawal, K. P., Garg, S., Sharma, S., & Patel, P. (2016). Development and validation of OPTICS based spatio-temporal clustering technique. *Information Sciences*, 369, 388–401. https://doi.org/10.1016/j.ins.2016.06.048.

Ankerst, M., Breunig, M. M., Kriegel, H. P., & Sander, J. (1999). OPTICS: Ordering points to identify the clustering structure. *ACM Sigmod Record*, 28(2), 49–60.

Appel, K. & Haken, W. (1977). Every planar map is four colorable. Part I: Discharging. *Illinois Journal of Mathematics*, 21(3), 429–490. https://doi.org/10.1215/ijm/1256049011.

Jain, A. K. (2010). Data clustering: 50 years beyond K-means. *Pattern Recognition Letters*, 31(8), 651–666.

Caliński, T. & Harabasz, J. (1974). A dendrite method for cluster analysis. *Communications in Statistics*, 3(1), 1–27. https://doi.org/10.1080/03610927408827101.

Duque, J. C., Ramos, R., & Surinach, J. (2007). Supervised regionalization methods: A survey. *International Regional Science Review*, 30(3), 195–220.

Assunção, R. M., Neves, M. C., Câmara, G., & da Costa Freitas, C. (2006). Efficient regionalization techniques for socio-economic geographical units using minimum spanning trees. *International Journal of Geographical Information Science*, 20(7), 797–811.

Coley, D. A. (1999). *An introduction to genetic algorithms for scientists and engineers*. World Scientific.

Lorena, L. A. N., & Furtado, J. C. (2001). Constructive genetic algorithm for clustering problems. *Evolutionary Computation*, 9(3), 309-327.

Chapter 4: Statistical cluster analysis

Getis, A., & Ord, J. K. (1992). The analysis of spatial association by use of distance statistics. *Geographical Analysis*, 24(3), 189–206.

Ord, J. K., & Getis, A. (1995). Local spatial autocorrelation statistics: Distributional issues and an application. *Geographical Analysis*, 27(4), 286–306.

Anselin, L. (1995). Local indicators of spatial association—LISA. *Geographical Analysis*, 27(2), 93–115.

Wong, D. (2009). The modifiable areal unit problem (MAUP). In Fotheringham, A. S., & Rogerson, P. A. (Eds.). (2008). *The SAGE handbook of spatial analysis*. (pp. 105–124). Sage.

Rice, J. A. (2003). *Mathematical statistics and data analysis*. China Machine Press.

Wasserstein, R. L., & Lazar, N. A. (2016). The ASA statement on p-values: Context, process, and purpose. *The American Statistician*, 70(2), 129-133. https://doi.org/10.1080/00031305.2016.1154108.

Benjamin, D. J., Berger, J. O., Johannesson, M., Nosek, B. A., Wagenmakers, E. J., Berk, R., & Johnson, V. E. (2018). Redefine statistical significance. *Nature Human Behaviour*, 2, 6–10. https://doi.org/10.1038/s41562-017-0189-z.

Chapter 5: Spatiotemporal pattern mining

Cheng, T., & Adepeju, M. (2014). Modifiable Temporal Unit Problem (MTUP) and its effect on space-time cluster detection. *PLoS ONE*, 9(6), e100465.

Ord, J. K., & Getis, A. (1995). Local spatial autocorrelation statistics: Distributional issues and an application. *Geographical Analysis*, 27(4), 286–306.

Anselin, L. (1995). Local indicators of spatial association—LISA. *Geographical Analysis*, 27(2), 93–115.

Tobler, W. (2014, December 10). Personal Communication. Esri/University of Redlands Colloquium. Redlands, California.

Chapter 6: Modeling spatial relationships and making predictions

Stigler, S. M. (1981). Gauss and the invention of least squares. *The Annals of Statistics*, 465–474.

Fox, J. (1991). *Regression diagnostics*. Sage. https://dx.doi.org/10.4135/9781412985604.

Brunsdon, C., Fotheringham, A. S., & Charlton, M. E. (1996). Geographically weighted regression: A method for exploring spatial nonstationarity. *Geographical Analysis*, 28(4), 281–298.

Fotheringham, A. S., Brunsdon, C., & Charlton, M. (2003). *Geographically weighted regression: The analysis of spatially varying relationships*. John Wiley & Sons.

Rice, J. A. (2003). *Mathematical statistics and data analysis*. China Machine Press.